T0214896

Communications in Computer and Information Science 568

Commenced Publication in 2007
Founding and Former Series Editors:
Alfredo Cuzzocrea, Dominik Ślęzak, and Xiaokang Yang

Editorial Board

Simone Diniz Junqueira Barbosa
*Pontifical Catholic University of Rio de Janeiro (PUC-Rio),
Rio de Janeiro, Brazil*

Phoebe Chen
La Trobe University, Melbourne, Australia

Xiaoyong Du
Renmin University of China, Beijing, China

Joaquim Filipe
Polytechnic Institute of Setúbal, Setúbal, Portugal

Orhun Kara
TÜBİTAK BİLGEM and Middle East Technical University, Ankara, Turkey

Igor Kotenko
*St. Petersburg Institute for Informatics and Automation of the Russian
Academy of Sciences, St. Petersburg, Russia*

Ting Liu
Harbin Institute of Technology (HIT), Harbin, China

Krishna M. Sivalingam
Indian Institute of Technology Madras, Chennai, India

Takashi Washio
Osaka University, Osaka, Japan

More information about this series at http://www.springer.com/series/7899

Xichun Zhang · Maosong Sun
Zhenyu Wang · Xuanjing Huang (Eds.)

Social Media Processing

4th National Conference, SMP 2015
Guangzhou, China, November 16–17, 2015
Proceedings

 Springer

Editors
Xichun Zhang
South China University of Technology
Guangzhou
China

Zhenyu Wang
South China University of Technology
Guangzhou
China

Maosong Sun
Tsinghua University
Beijing
China

Xuanjing Huang
Fudan University
Shanghai
China

ISSN 1865-0929   ISSN 1865-0937 (electronic)
Communications in Computer and Information Science
ISBN 978-981-10-0079-9   ISBN 978-981-10-0080-5 (eBook)
DOI 10.1007/978-981-10-0080-5

Library of Congress Control Number: 2015954349

Springer Singapore Heidelberg New York Dordrecht London

© Springer Science+Business Media Singapore 2015
This work is subject to copyright. All rights are reserved by the Publisher, whether the whole or part of the material is concerned, specifically the rights of translation, reprinting, reuse of illustrations, recitation, broadcasting, reproduction on microfilms or in any other physical way, and transmission or information storage and retrieval, electronic adaptation, computer software, or by similar or dissimilar methodology now known or hereafter developed.
The use of general descriptive names, registered names, trademarks, service marks, etc. in this publication does not imply, even in the absence of a specific statement, that such names are exempt from the relevant protective laws and regulations and therefore free for general use.
The publisher, the authors and the editors are safe to assume that the advice and information in this book are believed to be true and accurate at the date of publication. Neither the publisher nor the authors or the editors give a warranty, express or implied, with respect to the material contained herein or for any errors or omissions that may have been made.

Printed on acid-free paper

Springer Science+Business Media Singapore Pte Ltd. is part of Springer Science+Business Media
(www.springer.com)

Preface

We are living in an increasingly networked world. People, information, and other entities are connected via the World Wide Web, e-mail networks, instant messaging networks, mobile communication networks, online social networks, internet of things, etc. These generate massive amounts of social data, which present great opportunities in understanding the science of user behavioral patterns and the structure of networks formed by people interactions. The Fourth National Conference on Social Media Processing (SMP) was held in Guangzhou, China, in 2015 for the purpose of promoting original research in mining social media and applications, bringing together experts from related fields such as natural language processing, data mining, information retrieval, and social science, and providing a leading forum in which to exchange research ideas and results in emergent social media processing problems.

The conference received 105 submissions, of which 53 were English submissions. All papers were peer reviewed by at least three members of the Program Committee (PC) composed of international experts in natural language processing, data mining, information retrieval, and social science. The PC together with the PC co-chairs worked very hard to select papers through a rigorous review process and via extensive discussion. The competition was very strong; only 14 papers were accepted as full papers and nine as short papers. The conference also featured invited speeches from outstanding researchers in social media processing and related areas (the list may be incomplete): Ho-fung Leung (The Chinese University of Hong Kong), Irwin King (The Chinese University of Hong Kong), Shuo Bai (Shanghai Stock Exchange), Ting Liu (Harbin University), Yucheng Liang (Sun Yat-sen University), Bin Ke (National University of Singapore), Chunlin Duan (South China University of Technology), Jonathan Zhu (City University of Hong Kong), Wei Yang (Tencent), Chunyu Lin (TRS), Kai Chen (HYLANDA), and Jiyang Liu (Gridsum).

Without the support of several funding agencies and industrial partners, the successful organization of SMP 2015 would not have been possible. Sponsorship was provided by the following companies, among others: Tencent, TRS, HYLANDA, Gridsum, iFLYTEK, Sungy Mobile, DATATANG, and ZhiweiData. We would also like to express our gratitude to the Steering Committee of the special group of Social Media Processing of the Chinese Information Processing Society for all their advice and the Organizing Committee for their dedicated efforts. Last but not least, we sincerely thank all the authors, presenters, and attendees who jointly contributed to the success of SMP 2015.

November 2015

Xichun Zhang
Maosong Sun
Zhenyu Wang
Xuanjing Huang

Organization

Steering Committee Chair

Li Sheng   Harbin Institute of Technology, China

Steering Committee Co-chair

Li Yuming   Beijing Language and Culture University, China

Steering Committee

Bai Shuo	Shanghai Stock Exchange, China
Huang Heyan	Beijing Institute of Technology, China
Li Xiaoming	Peking University, China
Ma Shaoping	Tsinghua University, China
Meng Xiaofeng	Renmin University of China
Nie Jianyun	University of Montreal, Canada
Shi Shuicai	TRS
Sun Maosong	Tsinghua University, China
Wang Feiyue	Institute of Automation, Chinese Academy of Sciences, China
Zhou Ming	Microsoft Research Asia

President of the Assembly

Zhang Xichun	South China University of Technology, China
Sun Maosong	Tsinghua University, China

Program Committee Chairs

Wang Zhenyu	South China University of Technology, China
Huang Xuanjing	Fudan University, China

Program Committee

Chen Zhumin	Shandong University, China
Cheng Xueqi	Institute of Computing Technology, Chinese Academy of Sciences, China
Feng Chong	Beijing Institute of Technology, China
Feng Shizheng	Renmin University of China
Fu Guohong	Heilongjiang University, China
Gao Yue	National University of Singapore

Guo Hanqi	Argonne National Laboratory, USA
Han Qilong	Harbin Engineering University, China
Hong Yu	Soochow University, China
Huang Xuanjing	Fudan University, China
Ji Zhong	Tianjin University, China
Jiang Shengyi	Guangdong University of Foreign Studies, China
Jiang Wei	Beijing University of Technology, China
Jiang Yu-Gang	Fudan University, China
Li Aiping	National University of Defense Technology, China
Li Bing	The University of International Business and Economics, China
Li Guoling	Tsinghua University, China
Li Juanzi	Tsinghua University, China
Li Shoushan	Soochow University, China
Liang Jie	The University of Technology Sydney, Australia
Lin Chen	Xiamen University, China
Lin Hongfei	Dalian University of Technology, China
Liu Dexi	Jiangxi University of Finance and Economics, China
Liu Kang	Institute of Automation, Chinese Academy of Sciences, China
Liu Lizhen	Capital Normal University, China
Liu Pengyuan	Beijing Language and Culture University, China
Liu Ting	Harbin Institute of Technology, China
Liu Yang	Shandong University, China
Liu Yiqun	Tsinghua University, China
Liu Zhiyuan	Tsinghua University, China
Lu Hong	Fudan University, China
Ma Jun	Shandong University, China
Mao Wenji	Institute of Automation, Chinese Academy of Sciences, China
Mo Tong	Peking University, China
Peng Tao	University of Illinois at Urbana-Champaign, USA
Qi Haoliang	Heilongjiang Institute of Technology, China
Qian Tieyun	Wuhan University, China
Qin Bing	Harbin Institute of Technology, China
Ruan Tong	East China University of Science and Technology, China
Sha Ying	Institute of Information Engineering, Chinese Academy of Sciences, China
Shen Hao	Communication University of China
Shen Huawei	Institute of Computing Technology, Chinese Academy of Sciences, China
Shen Yang	Tsinghua University, China
Shi Chuan	Beijing University of Posts and Telecommunications, China
Shi Hanxiao	Zhejiang Gongshang University, China
Song Guojie	Peking University, China
Song Wei	Capital Normal University, China
Sun Guanglu	Harbin University of Science and Technology, China
Tang Jie	Tsinghua University, China

Wang Bailing	Harbin Institute of Technology, China
Wang Bin	Institute of Information Engineering, Chinese Academy of Sciences, China
Wang Daling	Northeastern University, China
Wang Mingwen	Jiangxi Normal University, China
Wang Shuaiqiang	University of Jyväskylä, Finland
Wang Suge	Shanxi University, China
Wang Ting	National University of Defense Technology, China
Wang Ying	Jilin University, China
Wang Zhenyu	South China University of Technology, China
Wang Zuchao	Peking University, China
Wu Dayong	Institute of Computing Technology, Chinese Academy of Sciences, China
Xia Yunqing	Microsoft
Xiong Jinhua	Institute of Computing Technology, Chinese Academy of Sciences, China
Xu Ruifeng	Harbin Institute of Technology, China
Yang Yanwu	Huazhong University of Science and Technology, China
Yang Zhihao	Dalian University of Technology, China
Yao Tianfang	Shanghai Jiaotong University, China
Ying Lan	Guizhou Normal University, China
Yuan Xiaoru	Peking University, China
Zhan Weidong	Peking University, China
Zhang Chengzhi	Institute of Scientific and Technical Information of China
Zhang Guoqing	Institute of Computing Technology, Chinese Academy of Sciences, China
Zhang Huaping	Beijing Institute of Technology, China
Zhang Ming	Peking University, China
Zhang Qi	Fudan University, China
Zhang Shu	Fujitsu R&D Center
Zhang Yu	Harbin Institute of Technology, China
Zhang Yuejie	Fudan University, China
Zhao Dongyan	Peking University, China
Zhao Jun	Institute of Automation, Chinese Academy of Sciences, China
Zhao Shiqi	Baidu
Zhao Yanyan	Harbin Institute of Technology, China
Zheng Chen	Anhui University, China
Zheng Xiaoling	Chinese Academy of Sciences, China
Zhou Dong	Hunan University of Science and Technology, China

Organizing Committee Chairs

Ye Weixiong	South China University of Technology, China
Yang Xiaowei	South China University of Technology, China

Best Paper Award Committee Chair

Ma Jun   Shandong University, China

Workshop Chairs

Shen Hao   Communication University of China
Feng Shizheng   Renmin University of China
Qin Bing   Harbin Institute of Technology, China

Publicity Chairs

Zhang Huaping   Beijing Institute of Technology, China
Guo Fen   South China University of Technology, China

Financial Chair

Chen Dongxiu   South China University of Technology, China

Student Sponsorship Chair

Huang Han   South China University of Technology, China

Publication Chair

Gong Yeyun   Fudan University, China

Organizing Committee

Guo Fen   South China University of Technology, China
Huang Han   South China University of Technology, China
Lu Yeli   South China University of Technology, China
Xu Ke   South China University of Technology, China
Zhang Anding   South China University of Technology, China

Sponsors

The Fourth National Conference of Social Media Processing (SMP 2015) is committed to building a network of social media processing researchers, which is sponsored by the following organizations.

Diamond

Silver

Bronze

Contents

Personalized Microtopic Recommendation with Rich Information

Yang Li[1], Jing Jiang[2], Ting Liu[1]([✉]), and Xiaofei Sun[1]

[1] Reseach Center for Social Computing and Information Retrieval,
Harbin Institute of Technology, Harbin, China
{yli,tliu,xfsun}@ir.hit.edu.cn
[2] School of Information Systems, Singapore Management University,
Singapore City, Singapore
jingjiang@smu.edu.sg

Abstract. Sina Weibo allows users to create tags enclosed in a pair of # which are called microtopics. Each microtopic has a designate page, and can be directly visited and commented on. Microtopic recommendation can facilitate users to efficiently acquire information by summarizing trending online topics and feeding comments with high quality. However, it is non-trivial to recommend microtopics to the users of Sina Weibo to satisfy their information needs. In this paper, we focus on personalized microtopic recommendation. Collaborative filtering based methods only utilize the user adoption matrix, while content based methods only use textual information. However, both of them can not achieve satisfactory performance in real scenarios. Moreover, auxiliary information on social media provides great potential to improve the recommendation performance. Therefore, we propose a novel hierarchical Bayesian model integrating user adoption behaviors, user item content information, and rich contextual information into the same principled model. We experiment with different kinds of textual and contextual information from both user and microtopic sides on a real dataset. Experimental results show that our model significantly outperforms a few baseline methods.

Keywords: Microtopic recommendation · Collaborative filtering · Topic model

1 Introduction

While Twitter is the most popular microblogging service in most parts of the world, Sina Weibo serves the majority of Chinese users. Microtopic is a special feature in Sina Weibo. A microtopic is represented as a word or phrase inside a pair of #, like a hashtag in Twitter. However, different from hashtags in Twitter, each microtopic in Sina Weibo has its own designated page. A microtopic typically has a host, a short description and rich attributes such as category and location. Users are encouraged to directly comment on its designated page. In this way, microtopics can improve user experience and boost online interactions.

© Springer Science+Business Media Singapore 2015
X. Zhang et al. (Eds.): SMP 2015, CCIS 568, pp. 1–14, 2015.
DOI: 10.1007/978-981-10-0080-5_1

Overall, microtopics in Sina Weibo go far beyond common posts or hashtags. They are more like threads in discussion forums.

Microtopics cover a wide range of topics, including not only trending events such as #马航飞机失联# (*Malaysian Airlines flight missing*) and #世界杯# (*World Cup*) but also long standing topics such as #深夜食堂# (*late night dining*) and #睡前阅读# (*bedtime reading*). Microtopics have played an important role in summarizing online information and feeding comments with high quality. With the proliferation of microtopics, many users encounter the problem of information overload. It is important to help users easily browse microtopics and find those of their interest. On the other hand, it is indispensable for fresh users who want to quickly follow big events or hot topics.

We focus on personalized microtopic recommendation in this work. Standard collaborative filtering based methods [15,19] can be directly applied to learn users' hidden interests via users' microtopic adoption history, but they suffer from the cold start problem. Moreover, these models can not take advantage of rich content and contextual information. For example, for each microtopic, we have posts and comments on its designated page; for each user, we can also obtain her post history. Such content information can presumably characterize properties of microtopics or indicate users' personal interests. Furthermore, we observe that in Sina Weibo, both users and microtopics have additional attributes such as gender of users and categories of microtopics. Similar to texts, these types of contextual information are valuable in profiling users and characterizing microtopics, which can connect similar users or similar microtopics. Intuitively, a joint model integrating all these rich information could help improve the recommendation performance. However, it is still not clear how to build an effective hybrid model for our problem.

In this paper, we propose a hierarchical Bayesian model based on collaborative filtering and topic modeling, to seamlessly integrate user adoption behaviors, user/item textual and contextual information into the same principled model. Unlike existing hybrid recommendation methods [8,16], by deeply incorporating the content from users' post history and comments on microtopics, our joint model benefits in topic modeling and gives interpretable representations of users and microtopics. By integrating both user and microtopic attributes, our model makes users or microtopics sharing the same attribute to have similar vectors in the latent factor space. Since a zero entry in the user-microtopic adoption matrix does not necessarily indicate that the user is not interested in the microtopic, we use a ranking optimization criterion to model users' preferences [21]. Note that we focus on microtopic recommendation in this paper. However, our model is flexible enough to be applied in other applications such as question recommendation in community-based question answering (cQA) services where we can also obtain rich content and contextual information from users and questions.

Our paper makes the following contributions:

- To the best of our knowledge, we are the first to study Sina Weibo microtopic recommendation problem on a large real dataset.

- We propose a novel hierarchical Bayesian model, which can seamlessly integrate user adoption behaviors, user/item content and contextual information into the same principled model. Our proposed model can be applied in other recommendation scenarios where both user and item have rich information.
- Through empirical evaluation, we find that both content and contextual information can help the recommendation task, our model significantly outperforms the state-of-the-art methods.

2   Microtopic Recommendation Model

In this section, we present our model for microtopic recommendation and explain the reasons behind the design of our model. We would like to consider several factors when designing our model. First, based on the idea of collaborative filtering, given a user and a new microtopic, to predict whether this user will be interested in this microtopic, we would like to make use of this user's as well as other users' historical records of microtopic adoption. We thus use these publishing records as indicators of users' interests in microtopics. Next, there are rich textual contents associated with both users and items. For a microtopic, we have the set of posts published on its microtopic page. For a user, similarly, we have her posts from timeline. Furthermore, attributes such as a user's gender and location or a microtopic's category can presumably also be useful. Finally, because for a given user we care about the accuracies of the top-ranked microtopics for her, we treat it as a ranking problem where for each user we would like

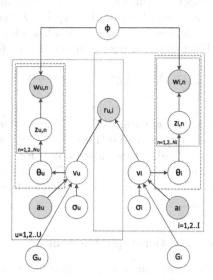

Fig. 1. Plate notation for our proposed microtopic recommendation model. The dashed rectangles are optional parts. Hyperparameters are omitted for clarity.

to rank the microtopics based on how likely the user is interested in browsing and commenting on a microtopic.

Our model mainly consists of three parts, namely, modeling user-microtopic adoptions, modeling user and microtopic content, and modeling user and microtopic attributes. Our overall model is illustrated in Fig. 1. In the rest of this section, we will present each of these three parts in detail. A basis of all three parts is that we assume there is a K-dimensional latent factor space. Each user and each microtopic is represented as a vector in this K-dimensional space. We use v_u, v_i to denote the vector for user u and microtopic i.

2.1   Modeling User-Microtopic Adoptions

The way we model microtopic adoption is similar to many existing latent factor models for recommendation. Given a user vector v_u and a microtopic vector v_i, we define an affinity score r_{ui} between user u and microtopic i as follows:

$$r_{ui} = v_u^\top v_i + b_u + b_i, \tag{1}$$

where r_{ui} models user u's preference to adopt microtopic i, and b_u and b_i are user bias and item bias to be learned.

Because microtopic recommendation belongs to the "one-class collaborative filtering problem" [20], where a zero entry in the adoption matrix indicates inaction rather than a negative rating, in order to rank items adopted by a user higher than items not adopted by her, we incorporated a ranking optimization criterion Bayesian Personalized Ranking (BPR) to learn user's preferences [21, 22]. Eq. (2) shows the objective function under the ranking criterion. Let r_{ui} denote the preference score computed from Eq. (1). Let $\mathcal{P}$ denote a set of triplets $\langle u, i, j \rangle$ derived from the training data where user u has adopted microtopic i but not microtopic j. The BPR criterion tries to minimize the following function:

$$\min_{\Theta} \sum_{\langle u,i,j \rangle \in \mathcal{P}} \ln(1 + e^{-(r_{ui}-r_{uj})}), \tag{2}$$

where Θ denotes the set of model parameters, i.e. the user and microtopic latent factor vectors and the bias terms. Here $\ln(1 + e^{-(r_{ui}-r_{uj})})$ can be considered the loss of ranking microtopic i higher than microtopic j for user u. The larger the value of $(r_{ui} - r_{uj})$ is, the smaller the loss is. Thus, the objective function is trying to maximize the difference between r_{ui} and r_{uj} when we know that i has been adopted but j has not.

2.2   Modeling User and Microtopic Content

To incorporate the content into our model, we first combine a user's (or microtopic's) posts into a pseudo document. We then use Latent Dirichlet Allocation (LDA) to model the generation of these pseudo documents. Inspired by recent work [18], we try to link LDA with collaborative filtering. Specifically, we assume that each of the K hidden factors that are used to represent users and micro-topics has a corresponding multinomial word distribution, denoted by ϕ_k, i.e. each hidden factor corresponds to a hidden topic in LDA. Each user (or micro-topic) has a distribution over the K topics, which is derived from its hidden factor vector. Let θ_u denote user u's topic distribution. We have

$$\theta_{u,k} = \frac{\exp(\kappa v_{u,k})}{\sum_{k'} \exp(\kappa v_{u,k'})}. \tag{3}$$

Similarly, a microtopic i's topic distribution θ_i can be derived from v_i.

Let w_u denote all the words in the pseudo document representing user u. Given the assumptions above, we can generate w_u using the following formula:

$$p(w_u \mid \theta_u, \phi) = \prod_n \sum_{z=1}^{K} \theta_{u,k} \phi_{k,w_{u,n}}. \tag{4}$$

The way the pseudo document for microtopic i is generated can be formulated similarly. With this, we can add the following term to the objective function (Eq. (2)) that needs to be minimized:

$$-\left(\sum_u \ln p(w_u|\Theta) + \sum_i \ln p(w_i|\Theta) \right), \tag{5}$$

where Θ denotes all the model parameters.

2.3 Modeling User and Microtopic Attributes

Finally, we would like to incorporate additional attributes that characterize users and microtopics into our model. Intuitively, users or microtopics sharing the same attribute are likely to have similar vectors in the latent factor space. To this end, we consider to embed a factor vector for each attribute value. Each user or microtopic is then profiled as an aggregation of the factor vectors of all its attributes.

Specifically, let a_u be an A_U-dimensional binary vector representing user u's attributes, where A_U is the total number of user attributes. $a_{u,t}$ is 1 if the attribute t is present in u and 0 otherwise. We then model user latent factors v_u as follows:

$$v_u = G_U^\top a_u + \sigma_u, \tag{6}$$

where $G_U \in \mathbb{R}^{A_U \times K}$ is a regression coefficient matrix and $\sigma_u \in \mathbb{R}^{K \times 1}$ is user u's deviation from the linear combination of the coefficients.

We profile each microtopic in a similar way. We pose zero-mean Gaussian priors on G_U, G_I, σ_u and σ_i.

2.4   Complete Model and Model Inference

We now present the complete model and model inference. In summary, we assume the following observations: For each user u, we observe a bag of words w_u and an attribute vector a_u. For each microtopic i, we observe also a bag of words w_i and an attribute vector a_i. We also have a set of triplets $\{\langle u, i, j \rangle\}$ indicating users' relative preferences between two microtopics. We have the following model parameters: For each latent factor (topic), there is a word distribution ϕ_k. For user attributes and microtopic attributes, there are two coefficient matrices G_U and G_I. Each user u has a user-specific latent factor vector σ_u and a bias term b_u. Similarly, each microtopic i also has a σ_i and a b_i. κ is the parameter which controls the transformation in Eq. (3). We use Θ to denote all model parameters. We further use $R(\Theta)$ to denote a regularization function on Θ derived from the prior distributions of all the model parameters. Recall that all model parameters have a zero-mean Gaussian prior except ϕ_k, which has a uniform Dirichlet prior.

The overall objective function we try to minimize is thus defined as follows:

$$\min_{\Theta} \sum_{\langle u,i,j \rangle \in \mathcal{P}} \ln(1 + e^{-(r_{ui} - r_{uj})}) - \mu \left(\sum_u \ln p(w_u | \Theta) + \sum_i \ln p(w_i | \Theta) \right) + \lambda R(\Theta).$$

We can see that the objective function includes three parts. The first part is the ranking optimization, the second part is the log likelihood of generating the textual content, and the last part poses regularization on all the parameters. μ and λ are manually defined scalar values to balance the relative contributions of each part.

To learn the model parameters, we use Monte Carlo EM [23], an inference method that alternates between collapsed Gibbs sampling and gradient descent. In the E-step, we fix all the parameters in Θ and compute θ_u and θ_i based on v_u and v_i, using Eqs. (3) and (6). We then collapse out the parameters ϕ to sample topic labels associated with each word. In the M-step, we fix the latent topic labels and perform gradient descent to learn parameters in Θ.

E-Step. In the E-step, we perform Gibbs sampling to learn the hidden variable $z_{u,n}$ by fixing all other parameters. In particular, we first compute θ_u from v_u. We then collapse out all the $\phi_{(\cdot)}$ and update each user u's n-th topic label as follows:

$$p(z_{u,n} = x \mid Z_{u,\neg n}, W, \theta_u, \beta) \propto \theta_{u,x} \cdot \frac{n^x_{w_{u,n}} + \beta}{n^x_{\cdot} + V\beta}, \tag{7}$$

where $n^x_{w_{u,n}}$ is the number of times topic x is assigned to word $w_{u,n}$, excluding the current word $w_{u,n}$'s topic assignment. V refers to vocabulary size, and β is the parameter of the Dirichlet prior on the $\phi_{(\cdot)}$.

M-Step. In this step, we perform gradient descent to learn the parameters by fixing the values of topic labels. We reformulate the objective function $\mathcal{L}$ as:

$$\mathcal{L} = \sum_{\langle u,i,j\rangle \in \mathcal{P}} \ln(1 + e^{-(r_{ui}-r_{uj})}) - \mu \sum_{u,n} \ln \theta_{u,z_{u,n}} - \mu \sum_{i,n} \ln \theta_{i,z_{i,n}} + \lambda R(\Theta).$$

Generally, for each parameter v, we compute the first derivative of $\mathcal{L}$ with respect to v as follows:

$$\frac{\partial \mathcal{L}}{\partial v} = \sum_{\langle u,i,j\rangle \in \mathcal{P}} \frac{\ln(1 + e^{-(r_{ui}-r_{uj})})}{\partial(r_{ui}-r_{uj})} \frac{\partial(r_{ui}-r_{uj})}{\partial v} - \mu \sum_{u,n} \frac{\ln \theta_{u,z_{u,n}}}{\partial v} - \mu \sum_{i,n} \frac{\ln \theta_{i,z_{i,n}}}{\partial v} + \lambda \frac{\partial R(\Theta)}{\partial v}.$$

In this case, each variable $v^{(t)}$ at iteration t can be updated by $v^{(t)} \leftarrow v^{(t-1)} - \frac{\partial \mathcal{L}}{\partial v}$. Our problem is resolved to compute $\frac{\partial(r_{ui}-r_{uj})}{\partial v}$, $\frac{\partial \ln \theta_{u,z_{u,n}}}{\partial v}$, $\frac{\partial \ln \theta_{i,z_{i,n}}}{\partial v}$ and $\frac{\partial R(\Theta)}{\partial v}$.

3   Experiments

In this section we empirically evaluate the various components of our proposed model for microtopic recommendation. We conduct experiments to answer the following research questions: (i) How much can collaborative filtering help for microtopic recommendation compared with a popularity-based baseline that is currently used by Sina Weibo? (ii) Does content help on top of collaborative filtering for this task, and if so, what content is the most useful? (iii) Does our method perform better than other baseline methods that also use a hybrid of collaborative filtering and content-based recommendation? (iv) Can user and microtopic attributes help the recommendation task, if so, which attributes are the most useful?

3.1   Data Set

Our data set is crawled from Sina Weibo. We started by selecting 100 seed microtopics published within three months before November 1st, 2014. We then crawled the users who had participated in these microtopics together with their comments published on the microtopics' pages. With the user name of these users, we were able to collect all the other microtopics which they had commented on. With these additional microtopics, we could repeat the same process for several times. All together we got 22,194 users and 164,462 microtopics. We removed those microtopics which had fewer than 5 participates. We also removed duplicate users and inactive users (with fewer than 30 followers or fewer than 50 posts). Finally we obtained 11,347 users, 13,188 microtopics and 783,118 posting records of these users on these microtopics. For the crawled users, we also obtained their profile information including gender, status (verified or unverified user) and location. For the microtopics, we crawled their category information.

3.2 Experimental Settings

Baseline Methods. For comparison, we consider the following baselines:

– **PR:** Popularity ranking. For each user, we rank microtopics based on the number of participants, which are obtained from Sina Weibo pages.
– **PMF:** Probabilistic matrix factorization [19]. The original model is designed for numerical ratings. For our task, we use 0s and 1s as rating scores.
– **BPR:** Bayesian personalized ranking matrix factorization [21]. BPR differs from PMF in that it offers an optimization criterion based on BPR for personalized ranking, which we adopted for our method.
– **OCCFWF:** In this method, user and item similarities based on content are tightly coupled with collaborative filtering by weighting the negative examples with similarity-based weights [16].
– **MCF:** Matrix Co-Factorization model proposed in [8], which incorporates rich content information and implicit feedback.

We refer to our proposed model as the Microtopic Recommendation Model (MTRM). Since we would like to empirically test the effectiveness of different sources of content, we first compare three degenerate versions of our model as follows. In all these three degenerate versions, no user or item attribute is incorporated yet.

– **MTRM-UC:** Our model incorporating users' posts as user content (i.e. a pseudo document for each user).
– **MTRM-IC:** Our model incorporating posts on microtopic pages as item content (i.e. a pseudo document for each microtopic).
– **MTRM-UCIC:** Our model incorporating both user's posts and posts on microtopic pages as content (i.e. a pseudo document for each user and a pseudo document for each microtopic).

Finally, as we will shown in Sect. 3.3, using user content is much more effective than item content for our problem. We then test the performance of our full model with user content and user/item attributes:

– **MTRM-UC-ATTR:** Our model incorporating user content and user/item attributes.

Experimental Setup. Similar to the settings of many other studies on recommendation [20,24], we hold out a percentage of the entries of the microtopic adoption matrix and use the remaining entries as training data. In particular, we perform 5-fold cross validation and we report the average performance. Recall that in BPR, we need to sample negative feedback to construct user preference data. For every user's each adopted microtopic in our training data, we randomly sample 5 microtopics that the user has not adopted as negative feedback. For other baseline methods, we use the same sampled data as negative instances for fair comparison. For evaluation, for each user in the test data, we randomly sample 1000 microtopics that the user has not adopted and have not been used as negative feedback. In other words, we make sure there is no overlap of user-microtopic pairs between the training and the test data.

For our models, we perform 200 runs of Monte Carlo EM. In each run, we run 10 iterations for Gibbs sampling and another 10 iterations of gradient descent. For the parameter μ which is balancing the likelihood of textual content and the adoption errors, we found that in MTRM-UC and MTRM-IC, when μ is set to between 0.01 and 0.1, we can achieve good performance. In MTRM-UCIC, we set the same $\mu = 0.01$ for both types of content. For all the zero-mean Gaussian priors in our model, we set the variances to be 0.01, and the regularization term λ is set to be 0.01 empirically. For MCF, we set the weight of negative instances to be 0.01 according to [8]. In OCCFWF, the weight is computed based on the content dissimilarity (Sect. 3.2). We tested with latent factor size K ranging from 10 to 50 with a gap of 10. Finally, we found for the baseline methods PMF, BPR and OCCWF, $K = 20$ is an optimal setting. In MCF and our models, $K = 30$ is an optimal setting. A larger K cannot improve the results.

We choose two evaluation metrics which are commonly used in one-class collaborative filtering tasks, namely Mean Percentage Ranking (MPR) [14] and Recall@M [24]. Due to the limited space, we omit the details of these metrics which can be found in [14,24]. Note that for MPR, the lower the value is, the better the results are.

3.3 Collaborative Filtering with Rich Content

Since our baseline methods do not make use of user/item attribute, we first compare the baselines with the versions of our model which do not use attribute information either. In other words, we compare the baselines with MTRM-UC, MTRM-IC and MTRM-UCIC. The goal here is threefold. First, we would like to see how much collaborative filtering can help the popularity-based baseline. Second, we would like to find out what content is useful for improving the recommendation results. Third, we would like to verify the performance of our method

Table 1. Comparison of PR, PMF, BPR, OCCFWF, MCF and three degenerate variations of our model MTRM. * indicates that the result is better than the method in the previous row at 5 % significance level by Wilcoxon signed-rank test.

Metric	MPR	Recall@10	Recall@50	Recall@100
PB	0.3381	0.0908	0.2091	0.3014
PMF	0.1252*	0.1677*	0.4552*	0.5997*
BPR	0.1178*	0.1725	0.4699	0.6077
OCCFWF	0.1169	0.1412	0.4294	0.5794
MCF	0.0984*	0.2351*	0.5084*	0.6460*
MTRM-IC	0.0945	0.2473*	0.4954	0.6318
MTRM-UC	**0.0822***	0.2729*	0.5227*	0.6529*
MTRM-UCIC	0.0829	**0.2830**	**0.5267**	**0.6590**

by comparing our degenerate models with other hybrid methods associating content features with collaborative filtering.

Result Analysis. Results in Table 1 shows the following: (i) PMF, the basic collaborative filtering method, clearly outperforms PR, the popularity-based method. The differences are quite substantial, showing that personalized recommendation of microtopics is very important. (ii) OCCFWF performs better than PMF in terms of MPR, but it gives low recalls. Among PMF, BPR and OCCFWF, BPR is giving consistent results in both MPR and recall, showing that for our microtopic recommendation task, a ranking based objective function gives more promising results than a rating based one. (iii) MCF and the three degenerate versions of MTRM are able to improve the recommendation performance over OCCFWF and BPR by deeply incorporating user-generated content into collaborative filtering. The results are consistent with previous findings in [8,13,18]. (iv) Comparing with MCF, our models always perform better in terms of both MPR and recall, although MCF has incorporated both user content and the microtopic content through matrix co-factorization. (v) Finally, we find that interestingly using pseudo documents for users is more effective than using pseudo documents for microtopics. We hypothesize that this is because the posts published on a microtopic's page are very diverse. In contrast, normal posts published by the same user may be more coherent and focused. Generally, we also find that the topics learned by MTRM-UC are more meaningful. The topics learned by MTRM-IC contain many meaningless words like "good", "applaud". When microtopics' pseudo documents are used on top of users' pseudo documents, the performance is very close to that of not adding them. Therefore, for the next experiment of using user/item attributes (Sect. 3.4), we use user content only.

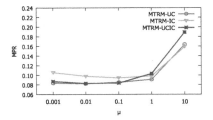 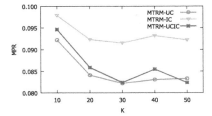

Fig. 2. MPR values with respect to different μ when $K = 30$.

Fig. 3. MPR values with respect to different K.

Parameter Sensitive Analysis. We would like to analyze how sensitive the performance of our model is with regard to the parameters. First, we vary the value of μ while fixing the other parameters. We show the results in terms of MPR for the three methods MTRM-IC, MTRM-UC and MTRM-UCIC in Fig. 2. Recall that μ controls the relative importance of collaborative filtering and content in the objective function. Figure 2 shows that the best results are achieved when μ is set to be between 0.01 and 0.1. The values of MPR increase when μ is larger than 1. Figure 3 shows the MPR results when we vary the number of topics K from 10 to 50. We use the following setting of μ: MTRM-UC ($\mu = 0.01$), MTRM-IC ($\mu = 0.1$) and MTRM-UCIC ($\mu_{UC} = \mu_{IC} = 0.01$). We find for all these three methods, the performance improves when K increaces. The result of MPR become flattened when K reaches 30. Through the overall results, in most settings, MTRM-UC and MTRM-UCIC perform much better than MTRM-IC, and MTRM-UC is close to MTRM-UCIC, meaning that modeling user content is empirically better than modeling microtopic content in our task.

3.4 Integrating Attributes

In this section, we empirically study how much user and microtopic attributes may help improve the recommendation results. Recall that in our model, we assume that the latent factor vector of a user or a microtopic is close to the linear combination of a set of coefficients corresponding to the attributes the user or microtopic has. In Sect. 3.3, we found that compared with MTRM-UC, MTRM-UCIC improves recall slightly but gives a much lower MPR, which means integrating user posts as content can capture most of the textual information. Next, we will incorporate user or microtopic attributes on top of the MTRM-UC model.

To incorporate users' attributes, we have collected their gender, verified status and location information. The results of incorporating users' attributes are shown in Table 2, indicated by +gender, +status and +location. We can see that adding gender information and location information turns out to be more useful in improving microtopic recommendation. A close examination of the data gives some examples. For instance, #我和闺蜜的那些事# is a microtopic for girls to share secrets they had with their girlfriends. Clearly this microtopic is meant for female users mostly. Another microtopic #养生美容知识# talks about cosmetology, which is also a female-oriented microtopic. We found that indeed very few male users would touch these microtopics. As for the attribute of verifica-

Table 2. Comparison of the results before and after incorporating each type of user and microtipic attributes. MTRM-UC-ATTR refers to our model using user content and best setting of features (user gender, user location and microtopic category).

Metric	MPR	Recall@10	Recall@50	Recall@100
MTRM-UC	0.0822	0.2729	0.5227	0.6529
+gender	0.0809*	0.2774	0.5419*	0.6723*
+status	0.0840*	0.2715	0.5359	0.6648
+location	0.0817	0.2773	0.5376*	0.6679*
+category	0.0792*	0.2810*	0.5421*	0.6728*
MTRM-UC-ATTR	**0.0789***	**0.2924***	**0.5474***	**0.6798***

tion status, since only around 10 % users are verified users, incorporating this attribute does not seem to be useful.

Furthermore, Weibo organizes microtopics in 16 main categories. We try to incorporate the category information of microtopics into our model to see if people's participation behaviors have some correlations with the categories of microtopics. In Table 2, we find that compared to MTRM-UC, the integration of category information (+category) gives more than 3.5 % decrease in MPR, and 3 % improvement in recall relatively. Compared to all the user attributes, microtopic category information is more useful. Finally, if we combine the attributes from users and microtopics, we find the best result we can achieve is using user gender, user location and microtopic category information (MTRM-UC-ATTR), as shown in Table 2.

4   Related Work

Collaborative Filtering: Recommendation methods can be classified as content-based recommendation [2,11], collaborative filtering [10,15], and hybrid approaches [5,16]. Many state-of-the-art hybrid approaches such as Matrix Factorization with Features [16], Matrix Co-Factorization Models [8,13] and Regression based Latent Factor Models [1,4] tried to combine content-based and CF approaches, however, these methods need efforts to collect and extract knowledge from item or user content. Wang and Blei [24] first applied Latent Dirichlet Allocation method (LDA) [3] on item specific textual content to recommend scientific articles. The method profiles each item as a combination of its topic distribution and a latent vector. In sentiment analysis of product reviews, [6] and [18] assume the topic distribution of each review is produced by the latent factors of the item. These methods could provide an interpretation to each latent factor because factors and topics are in the same space. Our method benefits from such approaches, but differently, we have textual content associated with both users and items (microtopics) in our task. Besides that, our model also combines auxiliary information on Sina Weibo.

Recommendation on Microblogs: With the popularity of microblogs, a growing number of studies have been proposed to provide better recommendation services. There are three main recommendation tasks involved, namely followee recommendation [11,12], tweet recommendation [4], and hashtag recommendation. Maybe the most related task to ours is hashtag recommendation, however, most work focuses on suggesting hashtag for a specific tweet. Godin et al. [9] uses topic models for Twitter hashtag recommendation. Ma et al. [17] proposed two PLSA-style topic models to incorporate user, time, hashtag, and tweet content for the task. Ding et al. [7] used topical translation model for hashtag suggestion on Sina Weibo. We study the task of recommending microtopics in Sina Weibo. To the best of our knowledge, this is the first work for this problem on a large real dataset.

5 Conclusions

In this paper, we study personalized microtopic recommendation in Sina Weibo. we propose a hierarchical Bayesian model to seamlessly integrate user adoption behaviors, user/item textual and contextual information into the same principled model. We design experiments to quantitatively evaluate our joint model against several state-of-the-art methods. We found it beneficial to incorporate users' historical posts to help learn better users' latent factor vectors. Furthermore, by incorporating both user and microtopic attributes, our model can further improve the recommendation performance.

Acknowledgements. We thank the anonymous reviewers for their constructive comments, and gratefully acknowledge the support of the National Basic Research Program (973 Program) of China via Grant 2014CB340503, the National Natural Science Foundation of China (NSFC) via Grant 61133012 and 61472107. We thank Minghui Qiu for helping to improve the work.

References

1. Agarwal, D., Chen, B.C.: Regression-based latent factor models. In: KDD (2009)
2. Balabanović, M., Shoham, Y.: Fab: content-based, collaborative recommendation. Commun. ACM **40**(3), 66–72 (1997)
3. Blei, D.M., Ng, A.Y., Jordan, M.I.: Latent dirichlet allocation. JMLR **3**, 993–1022 (2003)
4. Chen, K., Chen, T., Zheng, G., Jin, O., Yao, E., Yu, Y.: Collaborative personalized tweet recommendation. In: SIGIR (2012)
5. Claypool, M., Gokhale, A., Miranda, T., Murnikov, P., Netes, D., Sartin, M.: Combining content-based and collaborative filters in an online newspaper. In: SIGIR, vol. 60 (1999)
6. Diao, Q., Qiu, M., Wu, C.Y., Smola, A.J., Jiang, J., Wang, C.: Jointly modeling aspects, ratings and sentiments for movie recommendation (JMARS). In: KDD (2014)
7. Ding, Z., Qiu, X., Zhang, Q., Huang, X.: Learning topical translation model for microblog hashtag suggestion. In: IJCAI (2013)
8. Fang, Y., Si, L.: Matrix co-factorization for recommendation with rich side information and implicit feedback. In: HetRec (2011)
9. Godin, F., Slavkovikj, V., De Neve, W., Schrauwen, B., Van de Walle, R.: Using topic models for twitter hashtag recommendation. In: WWW (2013)
10. Goldberg, D., Nichols, D., Oki, B.M., Terry, D.: Using collaborative filtering to weave an information tapestry. Commun. ACM **35**(12), 61–70 (1992)
11. Hannon, J., Bennett, M., Smyth, B.: Recommending twitter users to follow using content and collaborative filtering approaches. In: RecSys (2010)
12. Hannon, J., McCarthy, K., Smyth, B.: Finding useful users on twitter: twittomender the followee recommender. In: Clough, P., Foley, C., Gurrin, C., Jones, G.J.F., Kraaij, W., Lee, H., Mudoch, V. (eds.) ECIR 2011. LNCS, vol. 6611, pp. 784–787. Springer, Heidelberg (2011)
13. Hong, L., Doumith, A.S., Davison, B.D.: Co-factorization machines: modeling user interests and predicting individual decisions in twitter. In: WSDM (2013)

14. Hu, Y., Koren, Y., Volinsky, C.: Collaborative filtering for implicit feedback datasets. In: ICDM (2008)
15. Koren, Y., Bell, R., Volinsky, C.: Matrix factorization techniques for recommender systems. Computer **42**(8), 30–37 (2009)
16. Li, Y., Hu, J., Zhai, C., Chen, Y.: Improving one-class collaborative filtering by incorporating rich user information. In: CIKM (2010)
17. Ma, Z., Sun, A., Yuan, Q., Cong, G.: Tagging your tweets: a probabilistic modeling of hashtag annotation in twitter. In: CIKM (2014)
18. McAuley, J., Leskovec, J.: Hidden factors and hidden topics: understanding rating dimensions with review text. In: Recsys (2013)
19. Mnih, A., Salakhutdinov, R.: Probabilistic matrix factorization. In: Advances in Neural Information Processing Systems, pp. 1257–1264 (2007)
20. Pan, R., Zhou, Y., Cao, B., Liu, N.N., Lukose, R., Scholz, M., Yang, Q.: One-class collaborative filtering. In: ICDM (2008)
21. Rendle, S., Freudenthaler, C., Gantner, Z., Schmidt-Thieme, L.: BPR: bayesian personalized ranking from implicit feedback. In: UAI (2009)
22. Shmueli, E., Kagian, A., Koren, Y., Lempel, R.: Care to comment?: recommendations for commenting on news stories. In: WWW (2012)
23. Wallach, H.M.: Topic modeling: beyond bag-of-words. In: ICML (2006)
24. Wang, C., Blei, D.M.: Collaborative topic modeling for recommending scientific articles. In: KDD (2011)

PRISM: Profession Identification in Social Media with Personal Information and Community Structure

Cunchao Tu, Zhiyuan Liu$^{(\boxtimes)}$, and Maosong Sun

State Key Lab on Intelligent Technology and Systems, National Lab for Information
Science and Technology, Department of Computer Science and Technology,
Tsinghua University, Beijing 100084, China
{tucunchao,lzy.thu}@gmail.com, sms@tsinghua.edu.cn

Abstract. User profession plays an important role in commercial services such as personalized recommendation and targeted advertising. In practice, profession information is usually unavailable due to privacy and other reasons. In this paper, we explore the task of identifying user professions according to their behaviors in social media. The task confronts the following challenges which make it non-trivial: how to incorporate heterogeneous information of user behaviors, how to effectively utilize both labeled and unlabeled data, and how to exploit community structure. To address these challenges, we present a framework of **PR**ofession **I**dentification in **S**ocial **M**edia (PRISM). It takes advantages of both personal information and community structure of users in the following aspects: (1) We present a cascaded two-level classifier with heterogeneous personal features to measure the confidences of users belonging to different professions. (2) We present a multi-training process to take advantages of both labeled and unlabeled data to enhance classification performance. (3) We design a profession identification method synthetically considering the confidences from personal features and community structure. We collect a real-world dataset to conduct experiments, and experimental results demonstrate significant effectiveness of our method compared with other baseline methods.

1 Introduction

Social media services, such as microblogs, enable users to post messages, share information and communicate with each other in social networks. Besides, users may also contribute tags and short notes to describe themselves. The user generated content (UGC) reserves rich facts about users, including their personality traits and social attributes. Many aspects and attributes of users have been investigated based on social media data, from simple attributes such as gender and age [2], to more complicated ones such as personality [23], happiness [6] and political polarity [21].

Profession, which is founded upon specialized educational training and aims to supply service to others, is also a critical social attribute of people. Sociologists

© Springer Science+Business Media Singapore 2015
X. Zhang et al. (Eds.): SMP 2015, CCIS 568, pp. 15–27, 2015.
DOI: 10.1007/978-981-10-0080-5_2

have been fascinated with user professions for a long time. It is a crucial factor for many social processes and dynamics, such as social organization, social control and cohesion, differentiation and inequality, power and influence, self and social identity [24]. With the development of social media, profession has become an important research subject of modern sociology. Besides benefiting research in sociology, user professions also make great contributions to commercial services such as personalized recommendation and targeted advertising. Professions of most users in social media, however, are implicit or regarded as a privacy issue. Hence, it will be beneficial for both academia and industry to effectively predict user professions based on large-scale social media data. To the best of our knowledge, user profession has been less investigated as a subject for prediction in social media. The task is the focus of this paper.

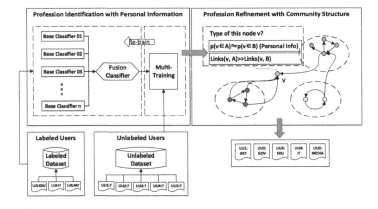

Fig. 1. The framework of PRISM.

The profession of a user is an essential part of human life. It may be explicitly or implicitly expressed in user generated content in social media. Hence, user professions can be identified according to user generated content. In this paper, we take microblogs as the representative social media, and explore the method of identifying user professions from microblog data.

In the context of microblog services, user professions are reflected in the following two aspects:

(1) *Personal Information.* Microblog users provide self descriptions and user tags, and constantly post short messages. The user generated contents form the personal information and can provide rich clues about user professions.
(2) *Network Information.* A user usually follows others to get information he/she is interested in. The following behaviors form social networks of microblog users. In our dataset we group users of the same professions into profession communities, which exhibit a relatively high modularity [20] score of 0.25. This indicates strong correlations between professions and network structure, and also confirms the homophily theory in sociology [17] that similar users tend to form social ties.

There are several challenges making profession identification non-trivial: (1) User generated personal information is heterogeneous. How can we integrate these information together for identification? (2) There are much more unlabeled users compared with those users labeled with professions. How can we effectively utilize both labeled and unlabeled data for identification? (3) Social networks also provide strong hints for user professions. How can we take advantages of community structure and further incorporate personal information together for identification?

To address these challenges, we propose an efficient framework of **PR**ofession **I**dentification in **S**ocial **M**edia (PRISM). PRISM takes advantages of both personal features and community structure, for user profession identification in social media.

Firstly, for heterogeneous personal information, we present a *cascaded two-level* classifier to measure confidences of users belonging to different professions. In the first level, we extensively extract features from different personal information sources, and build separate base classifiers for each source. Afterwards, a second-level classifier integrates the classification votes and makes final decision.Then, we further present a *multi-training* process, following the idea of co-training, to take advantages of both labeled and unlabeled users to improve classifier performance. Finally, we propose an profession identification method synthetically considering the confidences from personal features and community structure.

In the experiments, we collect more than 60 thousand manually annotated microblog users from Sina Weibo (http://weibo.com), the largest microblog service in China, as our dataset. According to characteristics of microblog users, we select 14 representative professions for study such as "art", "government", "sports" and "IT", etc. The experimental results on our dataset show that our method achieves the accuracy of 84.92 %, which outperforms all other baseline methods significantly.

2   The Framework of PRISM

We design the framework of our model as a two-step process: (1) We represent each user as multiple feature vectors extracted from various personal information sources, and build a cascaded two-level classifier to identify their professions. Furthermore, we introduce a multi-training process to improve classification performance by incorporating unlabeled data for training. (2) We further take advantages of profession community structure to refine profession identification. We introduce the details of our method as follows (Fig. 1).

2.1   Profession Identification with Personal Information

In this step, each user u is represented as a bag of feature vectors $\mathcal{X}_u = \{\mathbf{x}_{u,r}\}$. Here each $\mathbf{x}_{u,r}$ denotes a feature vector obtained from a distinct information source r, where $r \in \{1, \ldots, R\}$ and R is the number of information sources.

Suppose we have a set of annotated user-profession pairs $\{(\mathcal{X}_u, y_u)\}$ for training, where $y_u = k \in \{1, \ldots, K\}$ and K is the number of professions.

We build a cascaded two-level classifier for profession identification.

(1) **Base Classifier Construction.** For each information source r, we build a base classifier $f_r(\cdot)$ with a set of user-profession pairs $\{(\mathbf{x}_{u,r}, y_u)\}$. With these base classifiers, for a user u and its feature vector $\mathcal{X}_u$, we can obtain a identification matrix $\mathcal{P}_u = \{p_{k,r}\}$, where $p_{k,r} = \Pr(k|\mathbf{x}_{u,r}) = f_r(\mathbf{x}_{u,r}, k)$, indicating the confidence score for categorizing user u into profession k based on information source r.

(2) **Base Classifier Fusion.** We take identified results $\mathcal{P}_u$ obtained in (1) as input features, and construct a new set of user-profession pairs $\{(\mathcal{P}_u, y_u)\}$. Using these pairs, we build a fusion classifier $g(\cdot)$. The fusion classifier will assign a weight for each base classifier learned in Step 1, and fuse their identification results into the final identification scores, $\Pr(k|\mathcal{P}_u) = g(\mathcal{P}_u, k)$. We can then select the most confident label $\hat{y}_u = \mathrm{argmax}_k \Pr(k|\mathcal{P}_u)$, as the identified profession.

Feature Design and Base Classifier Construction. In social media, a user generates various types of content. Taking the user "Kai-Fu Lee", a famous Chinese IT activist, for example, he provides a short self description "CEO of Innovation Works", gives some user tags such as "venture capital", "innovation works", "education", "technology" and "e-business", and also has the verification information "Chairman and CEO of Innovation works". He also has posted thousands of messages, containing rich information including words, mentioned users, URLs, entities and hashtags. These information should be handled separately due to their distinct characteristics. In this paper, we consider eight distinct sources of user generated personal information to build features for base classifiers, which are listed in Table 1.

Table 1. Personal information sources.

No.	Name	Source description
1	DES	Self descriptions provided by user
2	TAG	User tags provided by user
3	VER	Verification information for user
4	MSG	Messages posted by user
5	MEN	Mentioned user IDs in messages
6	URL	URLs in messages
7	ENT	Named entities in messages
8	HAS	Hashtags in messages

Among these feature sources, the features in DES, VER and MSG are words extracted from text following the bag-of-word assumption. For TAG and HAS, we use tags as features. Besides using words in messages as features, we also extract user IDs identified by "@" in microblog messages as features of MEN, regard URLs in messages as features of URL which are usually in form of tiny URLs [1], and use named entity recognition (NER) tools to extract entities from messages as features of ENT.

For each feature source, there are tens of thousands of feature candidates. We have to perform feature selection to downsize feature sets. Following the valid experience in feature selection for text classification [10, 26], we use χ^2 statistic to select representative features for each feature source. Afterwards, we build base linear classifiers for each feature source.

Base Classifier Fusion. The prediction result $\mathcal{P}_u$ obtained from base classifiers for user u is a matrix, which can not be directly used as input of fusion classifier. We concatenate the transfer matrix $\mathcal{P}_u$ into a feature vector as input of fusion classifier, i.e., building a feature vector $\mathbf{z}_u$ simply by concatenating column vectors of $\mathcal{P}_u$, i.e., $z_{u,k+K\times(r-1)} = p_{k,r}$. The vector size of $\mathbf{z}_u$ is $K \times R$. We can also select the maximal scores or sum up scores of each row in the prediction matrix to build a feature vector. However, in experiments we find the concatenation scheme significantly outperforms the other schemes (max and sum), hence we only report the concatenation results.

We select Liblinear [7][1] to build base classifiers and fusion classifier. In this package, we select the method of L2-regularized logistic regression (LR), which is also the default setting of Liblinear. We have compared LR with SVM[2], and LR performs better in both effectiveness and efficiency. Hence, in the following part we only show the results obtained with LR.

Multi-training with Labeled and Unlabeled Data. In real world, there are much larger set of unlabeled users with no profession information. Here we want to employ the idea of co-training to perform multi-training of profession classification with both labeled and unlabeled data.

The basic idea is, after building base classifiers, we use them to identify professions for unlabeled users. We select the users that more than *half* base classifiers agree on their professions, and put these users with corresponding identified profession labels into training set. Then we re-train these base classifiers.

We can conduct the procedure iteratively until convergence. Multi-training is expected to enrich training data and improve classification performance with respect to both accuracy and generalization.

2.2 Profession Refinement with Community Structure

We observe from our dataset that users of the same professions tend to be friends and form communities in social networks, which is consistent with our intuition

---

[1] In this paper, we use the Java version of Liblinear, developed by Benedikt Waldvogel, which can be accessed via http://www.bwaldvogel.de/liblinear-java/.

[2] We select LibSVM [3] as the implementation of SVM, which can be accessed via http://www.csie.ntu.edu.tw/~cjlin/libsvm/.

and sociology theory [17]. Following [19], we assume that users of the same profession form an profession-specific community. A relatively high modularity [20] score of 0.25 obtained on our dataset confirms the assumption. Based on the observation, we take community structure into consideration to refine the identification results based on personal information.

Community-based profession refinement is formalized as follows. Suppose we have a social network $G = (U, E)$ and a subset of users who have profession labels and form communities for each profession, denoted as $G_k = (U_k, E_k)$ for profession k, where U_k is the set of all users of profession k and E_k is the set of edges between users in U_k. Afterwards, given a subset of users V with no profession labels, the task aims to extend existing communities by putting users from V into correct communities, i.e., assigning correct profession labels, according to the effect on community quality if users are involved in.

For community-based profession refinement, it is important to define an appropriate measure of community quality for each profession-based community G_k. The community quality can be verified from two aspects, including network structure and content information, i.e., *structure quality* and *content quality*.

Structure Quality. Structure quality measures the significance of a community from the perspective of network structure. It is intuitive that, the users with the same profession will form a dense and compact profession-based community.

To formally define structure quality, we give some definitions as follows. Take G_k, the community of profession k, for example, we define $U_{\neg k} = U \backslash U_k$. We also define $E_{k, \neg k}$ as the number of links between U_k and $U_{\neg k}$, and so do $E_{k,k}$ and $E_{\neg k, \neg k}$. We also have $E_k = E_{k,k} + E_{k, \neg k}$ and $E_{\neg k} = E_{\neg k, \neg k} + E_{\neg k, k}$.

Based on the above definitions, the structure quality of G_k is formalized as

$$Q_{structure}(G_k) = \frac{E_{k,k}}{E_{k,k} + E_{k, \neg k}} - \frac{E_k E_k}{E_k E_k + E_k E_{\neg k}}, \tag{1}$$

where the first entry indicates the proportion of how many links starting from U_k are connected within the community, and the second entry is that of the corresponding random graph. $Q_{structure}$ ranges $[-1, +1]$, and a strongly positive score indicates there is significant community structure in G_k.

This measure is originally proposed by [19] to compute the quality of a community, named as *normalized conductance*. In this paper, we integrate it with content quality together for profession refinement.

Content Quality. Content quality measures the significance of a community based on personal confidences of all users assigned in this community. In this paper, we employ identification confidences from our cascaded two-level classifier to measure content quality.

We define the content quality of a community as the average confidence scores over all users in this community, denoted as $Q_{content}(G_k)$. With content quality, the algorithm can take identification results based on personal information as input for refinement.

Profession Refinement. Afterwards, the overall community quality of G_k is defined as a combination of $Q_{structure}$ and $Q_{content}$,

$$Q(G_k) = \lambda Q_{structure}(G_k) + (1 - \lambda)Q_{content}(G_k), \tag{2}$$

where λ is a harmonic smoothing factor. When $\lambda = 1$, the quality measure is identical to that in [19].

With the measure $Q(\cdot)$, we conduct a greedy community extension as follows. Given a profession k, for each user $u \in V$ we compute

$$\Delta Q(u) = Q(G_k + u) - Q(G_k), \tag{3}$$

We find the user $\hat{u} = \arg\max_u \Delta Q(u)$, put $\hat{u}$ in U_k, and repeat the procedure until convergence.

After the community extension process, every unlabeled user is putted into an profession community, in which users have the most similar personal information and close connections. We take the types of matched communities as the final identified professions of unlabeled users.

3 Experiments and Analysis

We collect $62,415$ active and influential users from Sina Weibo. These users are all verified and categorized into 14 professions by officials of Sina Weibo, known as Hall of Fame in Weibo[3]. The ratios of various professions among these users are shown in Table 2. We also collect $150,000$ verified users with no profession annotations for multi-training.

From the profession composition of these labeled users, we find that the users in "media" and "government" are dominant. The reasons may be: (1) As the largest public social media service in China, public events are heavily discussed on Sina Weibo. Therefore, many people working in newspapers, news agencies and social media are active here. (2) The Government of China encourages their officials to go online and contact with citizens officially. Therefore, many national and local officials have registered in Sina Weibo.

3.1 Experimental Results on Profession Identification

We randomly divide the $62,415$ labeled users into training set and test set, of which $4/5$ is for training and $1/5$ for test. For the test set, we regard the labeled profession as the gold standard.

We select accuracy, macro-averaging precision/recall/F-Measure as evaluation metrics. Suppose the number of test users is U_{test}. If we get correct

---

[3] http://verified.weibo.com/.

Table 2. Ratios of professions in the annotated dataset. (%)

No.	Category	Ratio	No.	Category	Ratio
1	Media	25.6	8	Education	4.0
2	Government	15.1	9	Fashion	3.9
3	Entertainment	8.8	10	Games	3.8
4	Estate	8.2	11	Literature	3.4
5	Finance	7.0	12	services	3.4
6	IT	6.4	13	Art	3.1
7	Sports	5.6	14	Healthcare	1.7

identification for U_{correct} users, the accuracy is computed as $\frac{|U_{\text{correct}}|}{|U_{\text{test}}|}$. Accuracy evaluates per-user decisions across profession classes globally, and thus is micro-averaging. Whereas macro-averaging first calculates precision/recall for each profession class. That is, for profession k precision is $\frac{|U_{k,\text{correct}}|}{|U_{k,\text{predict}}|}$ and recall is $\frac{|U_{k,\text{correct}}|}{|U_k|}$, where U_k is the user set of profession k, and $U_{k,\text{predict}}$ is the user set that are predicted as profession k. And then it takes the average of these scores as overall precision P and recall R and further calculates F-measure as $\frac{2PR}{P+R}$.

Profession Identification with Personal Information. For feature selection of base classifiers, we evaluate performance with different numbers of features, and select $2,300$ features for DES, $3,800$ features for TAG, $4,000$ features for VER, 6600 features for MSG, 3200 features for MEN, 2700 features for URL, 3600 features for ENT and 4100 features for HAS, which achieve the best performance for each base classifier.

Table 3 shows the evaluation results on profession identification with various features of their combinations. In this table, the line of "Single Vector" is the baseline which represents a user by taking all features from multiple sources into a single vector, "Fusion" indicates the method of our cascaded two-level classifier, and "Fusion + MT" indicates the results after multi-training. From Table 3, we observe that:

(1) The fusion classifier performs much better than "Single Vector". This indicates that the design of cascaded two-level classifier is necessary and efficient for integrating heterogeneous feature sources.
(2) The base classifier using VER as feature source achieves the best performance among all base classifiers. This is consistent with the fact that verification descriptions are more informational and less noisy compared to other feature sources.
(3) The fusion classifier achieves much better performance compared to all base classifiers. This indicates that, the fusion of base classifiers can significantly improve identification.

(4) The accuracy and macro-averaging precision/recall/F-measure of fusion classifier with multi-training process are all larger than 80 %. This indicates the
identification capability of our classifier is balanced among various professions.

Table 3. Evaluation results for various features and combinations. (%)

Method	Accuracy	Precision	Recall	F
DES	31.25	51.82	28.90	37.11
TAG	38.11	50.55	31.04	38.46
VER	78.63	75.73	74.89	75.31
MSG	47.47	49.58	42.79	45.93
MEN	38.22	42.85	30.59	35.70
URL	26.38	36.47	13.68	19.90
ENT	33.86	36.88	26.95	31.15
HAS	30.91	37.44	17.60	23.94
Single vector	39.25	48.33	34.92	40.54
Fusion	81.25	79.60	76.27	77.90
Fusion+MT	**83.38**	**82.24**	**81.35**	**81.79**

Profession Refinement with Community Structure. To evaluate the performance of our profession refinement with community structure, we take two
community-based methods as baselines, i.e., label propagation algorithm (LPA)
and community detection (CD). LPA addresses the task as graph-based semi-
supervised learning [27]. The basic idea of LPA is the labels of a user is dependent
on its neighbors. By propagating labels from annotated users to unannounced
users through a social network, LPA can identify profession labels of users. As
previously introduced, CD is the user profiling algorithm proposed in [19]. The
both methods only consider community structure to classify users.

We show the evaluation results in Table 4. From the table we observe that:

(1) Profession refinement with community structure achieves considerable
improvement as compared to the two-level classifier. This indicates the community structure can also provide supplementary information for profession
identification beyond personal information.
(2) Profession refinement also outperforms two community-based baselines significantly. We can see that PRISM achieves the best result when $\lambda = 0.2$.
This indicates the effectiveness of personal information for profession identification. Here the community structure does not play a critical role for
profession refinement compared with personal information, because in Sina
Weibo personal information is much richer. The effect of social networks may
be emphasized in other scenarios with richer social structure information.

Table 4. Evaluation results of profession refinement with community structure. (%)

Method	Accuracy	Precision	Recall	F
LPA	58.86	57.05	54.53	55.76
CD	64.20	65.11	60.78	62.87
PRISM				
$\lambda = 0.1$	84.17	83.15	81.62	82.37
$\lambda = 0.2$	**84.92**	**83.78**	**81.89**	**82.82**
$\lambda = 0.3$	81.12	79.10	77.42	78.25
$\lambda = 0.5$	77.56	76.53	75.08	75.79

	1	2	3	4	5	6	7	8	9	10	11	12	13	14
1	91.05	0.66	1.88	0.81	1.05	1.43	0.51	0.30	0.24	0.30	0.39	0.95	0.42	0.03
2	2.64	93.45	0.17	0.39	0.84	0.28	0.03	0.78	0.02	0.00	0.05	0.17	0.39	0.78
3	9.72	0.47	81.34	0.19	1.43	0.66	0.28	0.19	2.09	0.57	1.15	0.38	1.53	0.00
4	5.31	0.40	0.26	82.76	5.82	1.04	0.40	0.26	0.51	0.00	0.26	2.71	0.26	0.00
5	3.95	2.22	0.29	3.56	77.37	6.55	0.29	1.92	0.77	0.29	0.29	2.03	0.39	0.09
6	7.78	0.44	0.00	2.52	9.42	72.18	0.33	0.88	0.55	4.06	0.11	1.43	0.11	0.22
7	3.45	0.36	0.36	0.36	0.79	0.00	94.05	0.24	0.00	0.12	0.00	4.00	0.24	0.00
8	3.89	3.05	0.85	0.68	4.40	1.19	0.34	77.82	0.34	0.00	0.85	3.72	1.19	1.69
9	5.93	0.18	4.14	0.54	2.88	0.90	0.36	0.72	81.31	1.26	0.18	0.90	0.72	0.00
10	3.19	0.00	1.60	0.18	0.00	5.14	0.71	0.00	1.60	86.70	0.35	0.18	0.35	0.00
11	9.33	1.87	1.65	0.00	1.45	0.00	0.00	2.28	0.20	0.83	78.65	0.00	3.32	0.42
12	13.23	1.04	1.86	4.34	6.41	4.76	0.00	3.72	1.04	0.83	0.00	62.56	0.21	0.00
13	4.71	2.35	4.71	0.00	2.35	0.00	0.79	2.95	0.79	0.79	3.33	0.20	77.05	0.00
14	2.64	0.76	0.37	0.00	2.64	0.37	0.00	1.13	0.75	0.00	1.51	0.00	0.01	89.82

Fig. 2. Distribution of identified professions for each profession.

To investigate the reason of identification errors, we show the distribution of identified professions for each profession in Fig. 2. In this figure, we define the entry of row-i and column-j as the ratio of the users in profession i being identified as profession j, i.e.,

$$e_{ij} = \frac{\sum_{u \in U_i} k_u = j}{|U_i|}, \tag{4}$$

where k_u indicates the identification result for the user u. To make the distribution comprehensive, we also illustrate the ratio in each entry using different shades of color. From this figure, we observe that:

(1) The profession "service" tends to be categorized into "media" by mistake, and the professions "art" and "education" are usually categorized into other professions incorrectly. The reason is that, there are more overlaps between these related professions, which makes the boundary between professions not so clear for identification. For example, the profession "service" usually interacts with "media" because they both involve in advertising and marketing.

(2) The professions "finance" and "IT" are usually categorized into each other incorrectly. We carry out extensive case studies and find that, many top

executives of companies usually have experiences in both "IT" and other business fields such as "finance", which cannot be well dealt with. The truth is also reflected in their friend network. In future, we may find more insight features to address these issues.

4   Related Work

User profiling aims to infer various attributes of users from social media [18]. These attributes can be roughly divided into explicit attributes (e.g., gender and age) and implicit attributes (e.g., interests, happiness and political orientation).

Existing user profiling studies mainly focus on explicit attributes, and usually adopt classification and recommendation methods for attribute prediction. Most classification-based works devote to extract efficient features from UGC to predict specific attributes, such as gender and age [2,9,12], location [15,21], tags [8,16] and other explicit labels [4,13,14].

Most explicit attributes are inferred from user-generated text data. For those attributes with rich sociality, social network structure may also be considered for prediction [15,19,22,25]. Researchers are also interested in implicate attributes such as personal interests [25], political orientation [21], personality traits [11,23] and social power [5].

This paper focuses on profession identification, which has been less studied by previous work. As compared with existing work, our framework compressively consider personal information, community structure and unlabeled data together to identify professions, which can be easily adapted to other social attributes of users.

5   Conclusion

This paper presents an efficient framework PRISM for profession identification in social media. The proposed PRISM identifies professions with both personal information and network structure, and addresses several practical challenging issues including incorporating heterogeneous information and utilizing unlabeled data. The experiments on a large real-world dataset demonstrate the effectiveness of PRISM, which can be easily extended to identify other attributes.

We plan to further explore the following research issues in the future: (1) This paper adopts a simple strategy, multi-training, to take advantages of unlabeled data. We will explore more sophisticated semi-supervised learning methods for profession identification. (2) Multiple social attributes of users may interact with each other and exhibit complicated correlations. We will explore joint identification of personal attributes such as age, gender, locations and professions. (3) Profession, as an important social attribute of people, will significantly influence people's many aspects such as language usage. We will extensively investigate these effects and patterns, which will be of great significance for both sociology research and commercial services.

Acknowledgement. This work is supported by the National Natural Science Foundation of China under Grant Nos. 61170196 and 61202140 and the Major Project of the National Social Science Foundation of China under Grant No. 13&ZD190.

References

1. Antoniades, D., Polakis, I., Kontaxis, G., Athanasopoulos, E., Ioannidis, S., Markatos, E.P., Karagiannis, T.: we.b: the web of short URLs. In: Proceedings of WWW, pp. 715–724 (2011)
2. Burger, J.D., Henderson, J., Kim, G., Zarrella, G.: Discriminating gender on twitter. In: Proceedings of EMNLP, pp. 1301–1309 (2011)
3. Chang, C.C., Lin, C.J.: LIBSVM: a library for support vector machines. ACM TIST **2**(3), 27 (2011)
4. Chaudhari, G., Avadhanula, V., Sarawagi, S.: A few good predictions: selective node labeling in a social network. In: Proceedings of WSDM, pp. 353–362 (2014)
5. Danescu-Niculescu-Mizil, C., Lee, L., Pang, B., Kleinberg, J.: Echoes of power: language effects and power differences in social interaction. In: Proceedings of WWW, pp. 699–708 (2012)
6. Dodds, P.S., Harris, K.D., Kloumann, I.M., Bliss, C.A., Danforth, C.M.: Temporal patterns of happiness and information in a global social network: hedonometrics and twitter. PLoS ONE **6**(12), e26752 (2011)
7. Fan, R.E., Chang, K.W., Hsieh, C.J., Wang, X.R., Lin, C.J.: Liblinear: a library for large linear classification. JMLR **9**, 1871–1874 (2008)
8. Feng, W., Wang, J.: Incorporating heterogeneous information for personalized tag recommendation in social tagging systems. In: Proceedings of KDD, pp. 1276–1284 (2012)
9. Fink, C., Kopecky, J., Morawski, M.: Inferring gender from the content of tweets: a region specific example. In: Proceedings of ICWSM (2012)
10. Forman, G.: An extensive empirical study of feature selection metrics for text classification. JMLR **3**, 1289–1305 (2003)
11. Golbeck, J., Robles, C., Turner, K.: Predicting personality with social media. In: Proceedings of CHI, pp. 253–262 (2011)
12. Goswami, S., Sarkar, S., Rustagi, M.: Stylometric analysis of bloggers' age and gender. In: Proceedings of ICWSM (2009)
13. Jacob, Y., Denoyer, L., Gallinari, P.: Learning latent representations of nodes for classifying in heterogeneous social networks. In: Proceedings WSDM, pp. 373–382 (2014)
14. Kong, X., Cao, B., Yu, P.S.: Multi-label classification by mining label and instance correlations from heterogeneous information networks. In: Proceedings of KDD, pp. 614–622 (2013)
15. Li, R., Wang, S., Deng, H., Wang, R., Chang, K.C.C.: Towards social user profiling: unified and discriminative influence model for inferring home locations. In: Proceedings of KDD, pp. 1023–1031 (2012)
16. Liu, Z., Tu, C., Sun, M.: Tag dispatch model with social network regularization for microblog user tag suggestion. In: Proceedings of COLING (2012)
17. McPherson, M., Smith-Lovin, L., Cook, J.M.: Birds of a feather: homophily in social networks. Ann. Rev. Sociol. **27**, 415–444 (2001)
18. Mislove, A., Lehmann, S., Ahn, Y.Y., Onnela, J.P., Rosenquist, J.N.: Understanding the demographics of twitter users. In: Proceedings of ICWSM (2011)

19. Mislove, A., Viswanath, B., Gummadi, K.P., Druschel, P.: You are who you know: inferring user profiles in online social networks. In: Proceedings of WSDM, pp. 251–260 (2010)

20. Newman, M.E.: Modularity and community structure in networks. PNAS **103**(23), 8577–8582 (2006)

21. Rao, D., Yarowsky, D., Shreevats, A., Gupta, M.: Classifying latent user attributes in twitter. In: Proceedings of Workshop on Search and Mining User-Generated Contents, pp. 37–44 (2010)

22. Sachan, M., Dubey, A., Srivastava, S., Xing, E.P., Hovy, E.: Spatial compactness meets topical consistency: jointly modeling links and content for community detection. In: Proceedings of WSDM, pp. 503–512 (2014)

23. Schwartz, H.A., Eichstaedt, J.C., Kern, M.L., Dziurzynski, L., Ramones, S.M., Agrawal, M., Shah, A., Kosinski, M., Stillwell, D., Seligman, M.E., et al.: Personality, gender, and age in the language of social media: the open-vocabulary approach. PLoS ONE **8**(9), e73791 (2013)

24. Volti, R.: An Introduction to the Sociology of Work and Occupations. Pine Forge Press, Thousand Oaks (2011)

25. Yang, S.H., Long, B., Smola, A., Sadagopan, N., Zheng, Z., Zha, H.: Like like alike: joint friendship and interest propagation in social networks. In: Proceedings of WWW, pp. 537–546 (2011)

26. Yang, Y., Pedersen, J.O.: A comparative study on feature selection in text categorization. Proc. ICML **97**, 412–420 (1997)

27. Zhu, X., Goldberg, A.B.: Introduction to semi-supervised learning. Synth. Lect. Artif. Intell. Mach. Learn. **3**(1), 1–130 (2009)

A Gaussian Copula Regression Model for Movie Box-office Revenue Prediction with Social Media

Junwen Duan, Xiao Ding, and Ting Liu[(✉)]

Reseach Center for Social Computing and Information Retrieval,
Harbin Institute of Technology, Harbin, China
{jwduan,xding,tliu}@ir.hit.edu.cn

Abstract. Previous work explored many kinds of features for the task of movie box-office prediction. However, little prior work has investigated the dependency relationships among these features. In this paper, we propose a novel Gaussian Copula regression model to study the correlation among predictive features. In particular, we first extract structured movie metadata and user activities on social media as features. We then apply Gaussian kernel to smooth out the data and learn the covariance matrix among the marginal distributions by maximum likelihood. We propose to approximately infer the movie box-office revenue by exploiting the covariance matrix. Experimental results show that our proposed method outperforms the baseline methods in the first week revenue prediction task and can achieve comparable performance on the gross revenue prediction task with a state-of-the art baseline in gross revenue prediction task. Our model is robust under various experimental settings.

Keywords: Copula regression · Movie revenue · Social media

1 Introduction

Predicting the revenues of up-coming movies is of clear interest to investors, movie related producers and movie theaters. Traditional approaches exploit structured metadata of a movie, such as its genre, MPAA rating, number of screens, to predict its future market performance, which show that movie box-office revenues are predictable.

However, this line of work suffers some limitations. On the one hand, they assumed that features used in the predication model are independent. Even though such restrictions make models easily scalable, they limit the expressiveness of models in some scenes. On the other hand, despite traditional models are good at capturing the linear and non-linear relationships, they still can not deal with arbitrary marginal distributions. Statistical analysis of historical movie market data show that movie relevant features follow different distributions [8]. For example, movie revenues are pareto law distributed, while number of theaters that a movie is shown follows a bimodal distribution.

© Springer Science+Business Media Singapore 2015
X. Zhang et al. (Eds.): SMP 2015, CCIS 568, pp. 28–37, 2015.
DOI: 10.1007/978-981-10-0080-5_3

To address these problems, we investigate a Gaussian Copula regression model to learn dependency relationships among features to predict movie box-office revenues. Given a movie, we automatically predict its first week and gross revenue by using both movie metadata and user activities on social media as features. The Copula we use in the paper is a family of distribution functions which is commonly used in statistical and economical domain. Even though Copula was first introduced in 1959 [7], it is a rather new topic in natural language processing and machine learning domain. It is capable of modeling multi-variate distribution by decoupling the multi-variate distribution to corresponding marginal distributions and correlation matrix. To the best of our knowledge, we are among the first to investigate Copula for the task of movie box-office revenue prediction.

To evaluate the performance of our approach, we construct a movie dataset with 188 movies. We compare our method with commonly used methods in regression tasks, including standard squared-loss linear regression, support vector regression with both linear kernel and RBF kernel and Gaussian process regression. The experimental results show that our approach outperforms the baseline methods in first week revenue prediction by a wide margin and can achieve comparable performance with state-of-the-art approaches in gross revenue prediction.

Experimental results of feature combinations indicate that user activities on social media are important indicators for predicating movie box-office performance. The main contributions of this paper are as follows.

- We are among the first to investigate Gaussian Copula regression model for movie box-office prediction task.
- We propose an approximate inference approach for Copula regression which is scalable.
- Our approach significantly outperforms the baselines in first week revenue prediction task.

In the rest of this paper, we summarize related work in Sect. 2. In Sect. 3, we first give a brief introduction to the theory of Copula and then describe our proposed model in detail. Experimental details are shown in Sect. 4. We discuss our findings in Sect. 5 and conclude this paper in Sect. 6.

2   Related Work

Movie box-office revenue prediction has attracted increasing attention in the area of natural language processing and machine learning. Traditional approaches exploit metadata of movies to tackle this problem. Sharda *et al.* [9] were among the first to study neural network model and movie metadata in the revenue prediction task. Instead of predicting the exact value, they formulated it as a classification problem by categorizing movie revenues into nice classes, ranging from "flop" to "blockbuster". In later work, Zhang *et al.* [12] applied back-propagation neural network to the task with similar settings.

Movie corresponding text is an alternative to movie metadata which requires high-level natural language processing techniques. Mishne *et al.* [6] analyzed bloggers' sentiment towards a movie and found that positive sentiment was a good predicator of movie success. Zhang *et al.* [13] applied sentiment analysis techniques to movie relevant news and achieved a fine performance by using both movie news and metadata. Joshi *et al.* [3] formulated the task as a text regression task by employing a linear regression model. They extracted text features such as n-grams and dependency relations from movie reviews.

User activities on social media have also been used in this task. Liu *et al.* [4] incorporated various activity features in their model, including sentiment polarity, number of relevant posts and user purchase intention. They achieved a relatively high performance compared to previous work.

Mestyán *et al.* [5] took a different prospective, they traced the user activities (create, edit, page view etc.) on the wikipedia page associated with a particular movie. They found strong correlations between the activities and gross revenues. By modeling these activities, they can predict the movie box-office revenue a month before its initial release.

3   Copula Regression for Prediction

Traditional machine learning methods like HMM are based on assumption of independent and identical distribution (i.i.d). They are only capable of modeling local dependency among random variables. Even though such restrictions make models easily scalable, in some sense they restrict the expressiveness of the models. Another disadvantage of these methods is that they often have to make prior assumptions about the underlying marginal distributions, they lack the ability to model margins with arbitrary distributions. However, our Copula regression take another prospective. On the one hand, we are able to model the dependency among variables explicitly by using a correlation matrix. On the other hand, it is able to deal with arbitrary distributions. Even with these settings, our model is still expressive. Since Copula is new to machine learning, we give a brief introduction to it.

3.1   A Brief Introduction to Copula

Copula was first introduced by Sklar in 1959 [7]. In statistical literature, it is well known as a family of distribution functions. The underlying mechanism of Copula is to separate joint distribution to uniform marginal distributions and a Copula function that connect them. Currently Copula families available include Clayton, Gumbel, Frank and so on. The idea of Copula can be summarized by Skalar Theorem [7].

Theorem 1 (Skalar's Theorem). *Suppose that there are n random variables $X_1, X_2, \cdots, X_n$. Let $F(X_1, X_2, \cdots, X_n)$ be their cumulative distribution function and $F_1(X_1), F_2(X_2), \cdots, F_n(X_n)$ be their corresponding marginal cumulative distribution functions. Then, if the marginal distributions are continuous, there exists a unique coupla C, such that*

$$F(x_1, x_2, \cdots, x_n) = C[F_1(x_1), F_2(x_2), \cdots, F_n(x_n)] \tag{1}$$

Note that, $F_1(x_1), \cdots F_n(x_n) \in [0, 1]$. Specially, C is independent of the marginals when the marginal distributions are continuous [2]. C can be considered as a function that maps $[0, 1]^n$ to $[0, 1]$. It is also true that with Copula function and marginal distributions of variables, we can model their joint distribution.

Copula makes no prior assumption about the prior distributions of marginals and it can deal with arbitrary marginal distributions. In this paper, we are using *Gaussian Copula*. The *Gaussian Copula* is defined as Eq. 2.

$$C_{Gaussian}(x_1, x_2, \cdots, x_n) = \Phi_{\Sigma}(\Phi^{-1}(x_1), \Phi^{-1}(x_2), \cdots, \Phi^{-1}(x_n)). \tag{2}$$

3.2 Kernel Density Estimation

Our training data is discrete and sparse (with only 188 samples), it overfits easily. We first apply non-parametric kernel density estimation to avoid overfitting issues and transform the discrete distributions to continuous ones. The kernel density estimation can be defined as Eq. 3:

$$\hat{f}_h(x) = \frac{1}{m} \sum_{i=1}^{m} K_h(x, x_i) \tag{3}$$

$K_h(\cdot)$ is the kernel function. In this paper, we employ Gaussian kernel (defined in Eq. 4) to smooth out the data. We use the default bandwidth in R package *"ks"*.

$$K_h(x, x_i) = \exp(-\frac{||x - x_i||_2^2}{2\sigma^2}) \tag{4}$$

Now that we have the estimated density, we can derive the cumulative distribution function by Eq. 5

$$\hat{F}(x) = \int_0^x \hat{f}_h(x) d_x. \tag{5}$$

3.3 Copula Parameter Estimation

With marginal distributions, we can construct the joint distribution among the stochastic variables by Copula function. In Eq. 6, $\hat{F}(x_i)$ are the smoothed covariates and $\hat{F}(y)$ is the smoothed response variable, in our paper, it is the movie box-office revenue. Φ_{Σ} is the standard multivariate Gaussian distribution with zero means and Σ variance. Σ^{-1} is the inverse CDF of standard Gaussian. All we need is to learn the Σ covariance matrix of the multi-variate Gaussian distribution. A commonly used approach is to apply maximum likelihood estimation. To make our model more robust, we follow Wang *et al.* [10] to calibrate the Σ. We find that the performance has slightly improvement when Σ is calibrated.

$$F(x_1, x_2, \cdots, x_n, y) = \Phi_{\Sigma}(\Phi^{-1}(\hat{F}_1(x_1)), \Phi^{-1}(\hat{F}_2(x_2)), \cdots, \Phi^{-1}(\hat{F}_n(x_n)), \Phi^{-1}(\hat{F}_y(y))). \tag{6}$$

3.4   Inference

To infer response value $F(y)$ from the multi-variate distribute, we have to calculate the mean response $E(F_y(y)|F_1(x_1), F_2(x_2)\cdots F_n(x_n); \Sigma)$. Wang *et al.* [10] prove that it is intractable in the multi-variate cases to obtain the exact inference. They propose to infer $F_y(y)$ by sampling the $F_y(y)$ to maximize the Gaussian Copula density. We take a different approach. Inspired by the random Copula function that generates data following a given Copula distribution, we use the covariance matrix, eigenvector and eigenvalues to perform the exact inference. Because Σ is symmetric and positive definite. It can be decomposed by Eq. 7, in which D is the eigenvector and A is the eigenvalues.

$$\Sigma = Ddiag(A)D^{-1} \tag{7}$$

$$R = (Ddiag(A^{\frac{1}{2}})D^T)^T \tag{8}$$

R is used to generate a particular distribution, it can be obtained by Eq. 8.

$$\hat{F}(y) = VR^{-1}_{[1:n,1:n-1]}R \tag{9}$$

$\hat{F}(y)$ is inferred by Eq. 9, in which $V = (\hat{F}_1(x_1), \hat{F}_2(x_2) \cdot \hat{F}_n(x_n))$. Note that $R_{[1:n,1:n-1]}$ is not a $n \times n$ matrix, we can not obtain its inverse directly. Thus, Moore–Penrose pseudo-inverse method is applied.

The idea of our algorithm can be summarized as follows.

1. Apply kernel density estimation to each of the features and revenues.
2. Obtain the marginal cumulative distribution and model the marginal distributions by Copula.
3. Estimate the parameter of the Copula by maximum likelihood estimation.
4. Infer the response values.

4   Experiments

The dataset for experiment consists of 188 movies between 2012 and 2014. Movies that have not arouse hot discussions on social media are excluded. All the experimental results described in this section are achieved by 5-fold cross validations.

4.1   Feature Sets

Our feature sets include three features i.e. *Number of screens, Purchase intention rate* and *Post rate*. Given a movie, we obtain *Number of screens* directly from public available data source. *Purchase intention rate* and *Post rate* require movie relevant microblogs in a particular time window, they are constructed following Liu *et al.* [4]. The details of each feature are as follows.

Number of Screens: The number of screens scheduled for the movie country-wide. The data is obtained from Wangpiao[1].

---

[1] Wangpiao: http://www.wangpiao.com.

Post-rate: Number of mircoblogs posted during a particular time window, which quantitatively measures the popularity of a particular movie on social media. We set the time window size to a week before the movie's official release.

$$Post - rate = \frac{|N_{total}|}{|time \quad window \quad size|} \tag{10}$$

Purchase Intention Rate: Number of movie relevant microblogs that have shown intention to see the movie in a particular time window. Bag-of-words, emoticon and length of microblog, url and mention features are used and SVM classification is applied to judge whether a microblog shows such intention.

4.2 Baselines

We employ commonly used baselines in regression tasks, including squared-loss linear regression, support vector regression (with linear kernel and RBF kernel) and Gaussian process regression (with RBF kernel). They are good at capturing the linearity and non-linearity of the features and prove to work well in several regression tasks, including movie box-office revenue prediction task [4]. Our approach is implemented with the R package *Coupla* [11]. The baselines are implemented with the Weka machine learning software(version 3.6) [1].

4.3 Evaluation Metrics

We employ coefficient of correlation (also known as R) and Relative Absolute Error (RAE) as our evaluation metric. R measures the correlation between predicted value and gold standard value, where 1 is positive correlated and 0 non-correlated. RAE (Relative absolute error) measures the prediction error, it is defined as,

$$RAE = \frac{\sum_{i=1}^{n} |\hat{y}_i - y_i|}{\sum_{i=1}^{n} |y_i - \bar{y}|} \tag{11}$$

where $\hat{y}_i$ is the predicted value, y_i is the exact value and $\bar{y}$ is the average of the extract values. They are both widely used as metrics in regression tasks.

4.4 Comparison to the Baselines

We compared our proposed Copula regression model to four baselines with all features. The detailed experiment result is shown in Table 1. Our proposed method outperforms the baselines in the first week revenue prediction task by a wide margin. It aligns with the best baseline SVR (linear kernel) in the gross revenue prediction task.

Table 1. The performance of our approach comparing to baselines in first week and gross revenue prediction

	First Week		Gross	
	R	RAE	R	RAE
Linear Regression	0.868	0.456	**0.787**	0.555
SVR (linear kernel)	0.861	0.425	0.777	0.496
SVR (RBF kernel)	0.858	0.542	0.773	0.554
GP Regression	0.803	0.451	0.723	0.530
Our Method	**0.874**	**0.367**	0.772	**0.489**

4.5 Varing the Amount of Training Data

To evaluate the sensitivity of our approach to train data, we sample 25 %, 50 %, 75 % data from our training set and evaluate it on all five models. The evaluation results are shown in Fig. 1. The left column shows the learning curves for coefficient of correlation. From the result, we find that our model is sensitive to training data size. Because we have to estimate the marginal distribution through kernel density estimation. With more training data, we can learn better distribution for the data. Our method quickly catches up when there is enough training data. The right column shows the learning curve for relative absolute error, we observed similar patterns.

4.6 Feature Combinations

We combine the features to validate whether they contribute to reduce the prediction error. The result is shown in Fig. 2. In the first week revenue prediction task, when we only use the metadata feature — number of screens, the *RAE* is over 50 %. When we add either *post rate* or *purchase intention rate* feature, the RAE drops dramatically. *Purchase intention rate* feature is slightly better than *post rate*. The performance reaches the best when all features are combined. Our experimental result is consistent with previous work that user activities on social media are alternatives to movie metadata in the task.

5 Discussion

Our proposed approach achieved a satisfying result in both first week and gross revenue prediction task. We are interested in why Copula works. The assumption of independent and identical distribution (i.i.d) is a strong restriction, which limits the expressiveness of models. However, our approach does not rely on such assumptions. Because Copula can decouple multivariate joint distributions while at the same time model the dependency among the random variables. Therefore, we can model marginal distributions independently.

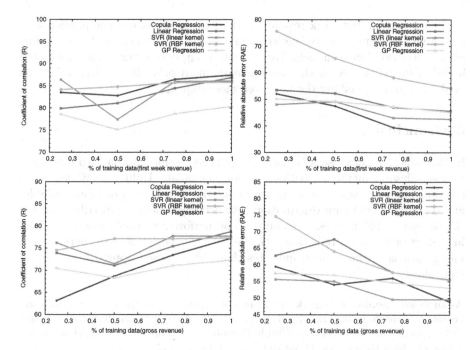

Fig. 1. Performance of our approach and baselines under different training data size. Left column: coefficient of correlation. Right column: Relative absolute error

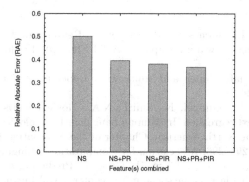

Fig. 2. Performance of our approach combining different features. *NS*: number of screens; *PR*: post rate; *PIR*: Purchase intention rate;

Another advantage we see from Copula regression is that it can model dependency of arbitrary distributions. The input of our model are the cumulative distributions of variables. Through methods like *probability integral transformation*, any forms of continuous distributions can be transformed to uniform distributions. The kernel density estimation in our model not only smooth out the data to avoid over-fitting issues, but also transform the data from discrete distribution to continuous one.

Comparing to first week revenue prediction, the gross revenue prediction is a relatively difficult task. We notice at first linear models achieve better performance than non-linear models. We are only using data several days before the initial movie release. After the movie release, many other factors that affect the prediction join in, making the task even more difficult.

6   Conclusion

In this paper, we propose a Gaussian Copula regression approach to model the relationships among user activities and movie metadata. Unlike traditional methods, our model makes no assumption about the distribution of input variables and is able to model joint distribution with arbitrary marginal distributions. On the dataset of 188 movies, our approach outperforms state-of-the-art baselines in first week revenue prediction task. We achieve consistent good performance under various settings which indicates that our proposed approach is effective and robust.

Acknowledgments. We thank the anonymous reviewers for their constructive comments, and gratefully acknowledge the support of the National Basic Research Program (973 Program) of China via Grant 2014CB340503, the National Natural Science Foundation of China (NSFC) via Grant 61133012 and 61472107.

References

1. Hall, M., Frank, E., Holmes, G., Pfahringer, B., Reutemann, P., Witten, I.H.: The weka data mining software: an update. ACM SIGKDD Explor. Newslett. **11**(1), 10–18 (2009)
2. Joe, H.: Multivariate Models and Multivariate Dependence Concepts. CRC Press, Boca Raton (1997)
3. Joshi, M., Das, D., Gimpel, K., Smith, N.A.: Movie reviews and revenues: an experiment in text regression. In: Human Language Technologies: The 2010 Annual Conference of the North American Chapter of the Association for Computational Linguistics, pp. 293–296. Association for Computational Linguistics (2010)
4. Liu, T., Ding, X., Chen, Y., Chen, H., Guo, M.: Predicting movie box-office revenues by exploiting large-scale social media content. Multimedia Tools Appl., 1–20 (2014)
5. Mestyán, M., Yasseri, T., Kertész, J.: Early prediction of movie box office success based on wikipedia activity big data. PloS One **8**(8), e71226 (2013)
6. Mishne, G., Glance, N.S.: Predicting movie sales from blogger sentiment. In: AAAI Spring Symposium: Computational Approaches to Analyzing Weblogs, pp. 155–158 (2006)
7. Nelsen, R.B.: An Introduction to Copulas, vol. 139. Springer Science & Business Media, Dordrecht (2013)
8. Pan, R.K., Sinha, S.: The statistical laws of popularity: universal properties of the box-office dynamics of motion pictures. New J. Phys. **12**(11), 115004 (2010)
9. Sharda, R., Delen, D.: Predicting box-office success of motion pictures with neural networks. Expert Syst. Appl. **30**(2), 243–254 (2006)

10. Wang, W.Y., Hua, Z.: A semiparametric gaussian copula regression model for predicting financial risks from earnings calls. In: Proceedings of the 52nd Annual Meeting on Association for Computational Linguistics (2014)
11. Yan, J., et al.: Enjoy the joy of copulas: with a package copula. J. Stat. Softw. **21**(4), 1–21 (2007)
12. Zhang, L., Luo, J., Yang, S.: Forecasting box office revenue of movies with bp neural network. Expert Syst. Appl. **36**(3), 6580–6587 (2009)
13. Zhang, W., Skiena, S.: Improving movie gross prediction through news analysis. In: Proceedings of the 2009 IEEE/WIC/ACM International Joint Conference on Web Intelligence and Intelligent Agent Technology, vol. 01, pp. 301–304. IEEE Computer Society (2009)

Personalized Hashtag Suggestion for Microblogs

Juan Xu, Qi Zhang[✉], and Xuanjing Huang

Shanghai Key Laboratory of Intelligent Information Processing,
School of Computer Science, Fudan University, Shanghai, China
{13916430285,qz,xjhuang}@fudan.edu.cn

Abstract. In microblogging services, users can generate *hashtags* to categorize their tweets. However, a majority of microblogs do not contain hashtags, which has intrigued active research on the problem of automatic hashtag recommendation for microblogs. Previous work conducted on this problem mostly does not take the user's preference into consideration. In this paper, we propose a novel personalized hashtag recommendation method for microblogs based on a probabilistic generative model which exploits users' perspectives on microblog posts for hashtag generation. Our experiments on a real microblogs dataset show that the proposed method outperforms state-of-the-art methods. We also show some case studies that demonstrate the advantages of considering both the content and user's personal preferences for hashtag suggestion.

1 Introduction

Microblogging services overload us with information, bombarding us with thousands of tweets, blog posts, and status updates every day. For example, Twitter, one of the most popular microblogging tools, has grown rapidly, with an estimated 200 million users generating 400 million tweets per day recently[1]. To cope with the volume of information shared daily, hashtags keywords prefaced with "#" in microblogging services have been introduced to help users categorize and search for tweets. Past empirical research shows that hashtags can be useful in many applications, including sentiment analysis [3,15], breaking event discoveries [2], query expansion [1], etc. In the spectrum of industry, Google started supporting Google+ hashtags in search queries on Sep 25, 2013[2]. Despite the availability and the usefulness of this feature, only 12.84 % tweets are marked with hashtags. Inclusion of hashtags in tweets is completely voluntary and user dependent. Thus, how to automatically generate or recommend hashtags has become an important research topic and drawn increased attention recently.

The task of hashtag recommendation is to automatically generate a short list of relevant hashtags as suggestions for a given tweet. Since it was first introduced by Mazzia and Juett [13], several methods have been proposed to tackle this problem [5,6,12,14]. Indeed, these studies are successfully tackling many

---

[1] https://blog.twitter.com/2013/celebrating-twitter7.
[2] http://techcrunch.com/2013/09/25/google-starts-supporting-google-hashtags-in-search-queries/.

© Springer Science+Business Media Singapore 2015
X. Zhang et al. (Eds.): SMP 2015, CCIS 568, pp. 38–50, 2015.
DOI: 10.1007/978-981-10-0080-5_4

notable challenges (i.e., hashtag sparseness, content shortness because of the 140-character limit, the vocabulary gap between tweets and hashtags, and topic diversity because of the open access in social media [18]) in hashtag recommendation. Nevertheless, they do not take into account personal preferences when recommending hashtags.

However, we believe that hashtag recommendation should be personalized. Users often utilize very different hashtags for their tweets [16]. For example, regarding *"The quarterfinal match between Roger Federer and Jo-Wilfried Tsonga in the French Open (a major tennis tournament held in Paris) 2013,"* tweets posted by users about this match may contain diverse tags for different purposes, such as "**#FrenchOpen13**," "**#RolandGarros1/4**" or "**#tenni**," which are used by those who just watched the game and posted a tweet about it, or "**#Roger, go**," "**#Roger, Allez & Come on**" or "**#Hero, Federer**" which are used by Roger's fans to represent their personal perspectives on this topic. Therefore, we would like to consider users' preferences in the choice of hashtag suggestions for tweets.

In this paper, we propose a personal-topic-translation model (PTTM) exploring users' perspectives on tweets to recommend personalized hashtags for microblogs. Our proposed method is a comprehensive generative model. We introduce a user perspective latent variable to take users preferences into consideration as well as exploiting topical perceptions about microblogs in hashtag suggestion. We evaluate our model on a real-world dataset with posts that have been assigned with hashtags. We find that compared with other models, hashtags recommended by our model are more accurate and less redundant within the top-ranked results. We also use some examples to explain the advantages of our model.

2 Related Works

Recently, increased efforts have been made to address the problem of hashtags recommendation for a certain microblog post in microblogging platforms. Mazzia and Juett [13] provide a preliminary suggestion system, using a Naive Bayes approach, with much focus on pre-processing steps. Further, some methods exploit to compute the similarity between tweets based different similarity metric, and then to recommend hashtags from similar tweets. Zangerle et al. [16] investigate three different approaches to recommend hashtags based on a TF-IDF representation of the tweet. They rank the hashtags based on the overall popularity of the tweet, the popularity within the most similar tweets, and the most similar tweets. Li et al. use Euclidean distance as the similarity metric to suggest hashtags from similar tweets [11]. Zangerle et al. [17] explore five text similarity functions as the similarity measure for the computation of recommendations.

However, the suggested tags are sparse. Therefore, a few works have been conducted to address this issue. They propose methods for general hashtag recommendation based on the underlying topics of the tweets. For example, Ding et al. [6] propose a topic-specific translation model, which regards hashtags and tweets as parallel description of a resource, and then combine topic model and word alignment model to recommend the hashtags. Our method is partly based on their study. But the striking difference is that we model the hashtag generation with user factor. So that the hashtags would suit both user's preferences and the theme of tweet content. Godin et al. [8] present an approach relies on Latent Dirichlet Allocation (LDA) to model the underlying topic assignment of tweets for general hashtags recommendation. But this approach have an inherit problem, that is the keyphrases extracted are too general to capture the tweet themes well.

All previous approaches always return the same list of tags for the same item regardless of user's preference. This problem is noted by [10]. Therefore, they propose a hashtag recommendation method based on collaborative filtering, which combines hashtags of similar users and similar tweets. TF-IDF approach is used to construct a feature vector for each tweet. Cosine similarity is used to compare the feature vectors. Although their approach considers user's preference, it ignores many other issues, such as tag sparse problem, etc. Therefore, in this paper, we propose a method attempting to address those challenges in hashtag recommendation problem.

3   Method

3.1   Preliminaries

We first introduce the notation used in this paper and formally formulate our problem. Let D to present an annotated corpus microblog posts, denoted as $d_1, d_2, \ldots, d_D$. Each post d_i is generated by a user u_i, where u_i is an index between 1 and U, and U is the total number of users. Each d_i consists of a pair of content words and assigned hashtags (w_i, t_i), where w_i and t_i are an index between 1 and D respectively. Each w_i contains a bag of words, denoted as $\{w_{i1}, w_{i2}, \ldots, w_{iN_i}\}$, where w_{iN_i} is an index between 1 and W, and W is the word vocabulary size. N_i is the number of words in w_i. Each t_i contains a bag of hastags, denoted as $\{t_{i1}, t_{i2}, \ldots, t_{iM_i}\}$, where t_{iM_i} is an index between 1 and T, and T is the hashtag vocabulary size. M_i is the number of hashtags in t_i. Given an unlabeled data set, the task of personalized hashtag recommendation is to discover a list of hashtags for each post with perceptive of both users' preferences and tweet themes.

3.2   Model Formulation

We first describe how we address the vocabulary gap between hashtags and microblogs, and the topic diversity issue. Topical word trigger model proposed by Liu et al. [12] and Ding et al. [6] have been shown to be effective for solving these two issues, in which they assume that hashtags and tweets as parallel description of a resource, and a document contains a mixture of topics, and each word has a hidden topic label. From this perspective, hashtag suggestion can be regarded as a translation process from a given post content to tags under a specific topic. While this assumption works well on long documents, for short microblog posts, posts are noisy and a single post tends to be about a single topic. Recently, there has been much progress in modeling topics for short texts [5], which assumes a single topic assignment for an entire tweet and also assumes a background word distribution ϕ^B that captures common words. Similar idea has also been used in the works of Zhao et al. [18] and Diao et al. [4]. Based on these works, we introduce a topic model which is pretty suitable for microblogs in our method.

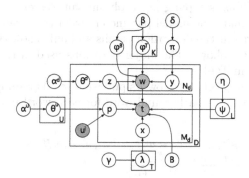

Fig. 1. The plate representation of the personal topical translation model.

As we discussed in Sect. 1, an important property of hashtag is that many hashtags are about user's personal perspectives on microblog posts rather than the themes of posts only. Thus our focus is to consider the impact of both microblog posts and user's preference for suggesting hashtags. To this end, an intuitive idea is that hashtags are either generated from posts or from user's perspectives. Therefore, we introduce a topic distribution θ^u for each user to capture her perspectives.

The proposed model is designed based on the following assumptions. When a user wants to write a tweet, she first generates the content, and then generates the hashtags. When she starts to write the content, she first chooses a topic based on the topic distribution θ^d. With the selected topic, words in the post are generated from the word distribution for that topic or from the background word distribution that captures white noise. During the generative process for hashtags, she first decides whether to tag about the post theme or her personal perspective. If she chooses the former, the hashtag is annotated according to the post topic. Otherwise, she selects a perspective according to her own perspective distribution θ^u to generate the hashtag. With the chosen generative source, hashtag t is either annotated according to the topic-dependent translation possibility $P(t_{dm}|w_d, z_d, \boldsymbol{B})$, where $P(t_{dm}|w_d, z_d, \boldsymbol{B}) = \sum_{n=1}^{N_d} p(t_{dm}|w_{dn}, z_d, \boldsymbol{B}) \cdot p(w_{dn}|w_d)$, and $\boldsymbol{B}$ presents the topic-specific word alignment table between a word and a hashtag which estimated by the combination of topic model and word alignment model, in which $B_{i,j,k} = P(t = t_j | w = w_i, z = k)$ is the word alignment probability between the word w_i and the hashtag t_j for topic k, or drawn from the tag distribution ψ^l of the perspective l. A variable x is introduced to decide the source of each hashtag, and we use λ to denote the probability of choosing to annotate according to the post theme rather than her personal perspective. Formally, the generation process of tweets is summarized in Fig. 2. The plate representation of the proposed model is depicted in Fig. 1.

3.3   Learning and Inference

We use collapsed Gibbs sampling [9] to obtain samples of the hidden variable assignment and estimate the model parameters from these samples. Due to the space limit, we leave out the derivation details and only show the derived Gibbs sampling formulas as follows. The major notations used in the following equations are explained in Table 1.

First, for the d-th tweet, we know its publisher u_d. The sampling probability of being a topic word or a background word for each word $v_i = w$ in dth tweet is sampled from:

$$p(y_{di} = s | v_i = w, z_{-i}, v_{-i}, y_{-di}, \delta, \beta) \propto$$
$$\frac{M_{s,-i} + \delta}{M_{.,-i} + 2\delta} \cdot \frac{M_{h,-i}^{w_{di}} + \beta}{M_{h,-i}^{(\cdot)} + W\beta}, \tag{1}$$

where $h = B$ when $s = 0$ and $h = z_d$ when $s = 1$. $M_{0,-i}$ and $M_{1,-i}$ are counters to record the numbers of words assigned to the background model and any topic, respectively. $M_{B,-i}^{w_{di}}$ is the times of w_{di} that assigned to background words. $M_{z_d,-i}^{w_{di}}$ is the numbers of w_{di} that are assigned to topic z_d. $-i$ indicates taking no account of the current position i.

Then we sample the tweet topic variable z_d for d-th tweet using:

$$p(z_d = k | w_d, z_{-d}, y, \alpha^d, \beta) \propto$$
$$\frac{M_{k,-d} + \alpha^d}{M_{.,-d} + K\alpha^d} \cdot \prod_{i=0}^{N_d} \frac{M_{k,-d}^{w_{di}} + \beta}{M_k^{(\cdot)} + W\beta}, \tag{2}$$

where $M_{k,-d}$ is the numbers of tweets that are assigned with topic k in the corpus; $M_{k,-d}^{w_{di}}$ is the number of occurrences of topic word w_{di} that is assigned with topic k, here topic word refers to word whose latent variable y equal 1; $-d$ indicates taking no account of the current tweet w_d.

We still have two latent variables which are tag topic (or tag perspective) variable p and the generative source of each hashtag variable x. We can jointly sample them based on the values of all other hidden variables. As we described in our model when the tag source variable $X = 1$, topic for each tag is generated from it's tweet topic and that has been sampled in formulas (2). Therefore, we just need to sample the user's perspective variable for each tag $q_j = t$ when the tag source variable $X = 0$:

$$p(x_{dj} = 0, p_{dj} = l | q_j = t, t_{-j}, p_{-j}, \gamma, \alpha^u, \eta) \propto$$
$$\frac{n_{t,-j} + \gamma}{n_{t,-j} + \tilde{n}_t + 2\gamma} \cdot \frac{M_{u,-j}^l + \alpha^u}{M_{u,-j}^{(\cdot)} + L\alpha^u} \cdot \frac{M_{l,-j}^t + \eta}{M_{l,-j}^{(\cdot)} + T\eta}, \tag{3}$$

---

1. Draw $\phi^B \sim$ Dirichlet (β), $\pi \sim$ Beta (δ), λ^t $\sim$ Beta (γ)
2. For each topic $k = 1, \ldots, K$
   (a) Draw $\phi^T \sim$ Dirichlet (β)
3. For each user $u = 1, \ldots, U$
   (a) Draw $\theta^u \sim$ Dirichlet (α^u)
4. For each perspective $l = 1, \ldots, L$
   (a) Draw $\psi^l \sim$ Dirichlet (η)
5. For each tweet $d = 1, \ldots, D$ created by $u = 1, \ldots, U$
   (a) Draw $\theta^d \sim$ Dirichlet (α^d)
   (b) Draw $z_d \sim$ Multinomial (θ^d)
   (c) for each word $n = 1, \ldots, N_d$
    i. Draw $y_{dn} \sim$ Bernoulli (π)
    ii. if $(y_{dn} = 1)$:
      Draw $w_{dn} \sim$ Multinomial (ϕ^{z_d})
    iii. if $(y_{dn} = 0)$:
      Draw $w_{dn} \sim$ Multinomial (ϕ^B)
   (d) for each hashtag $m = 1, \ldots, M_d$
    i. Draw flag $x_{dm} \sim$ Bernoulli $(\lambda^{t_{dm}})$
    ii. if $(x_{dm} = 1)$:
      Draw $t_{dm} \sim P(t_{dm}|w_d, z_d, \mathbf{B})$
    iii. if $(x_{dm} = 0)$:
      Draw $p_{dm} \sim$ Multinomial (θ^u)
      Draw $t_{dm} \sim$ Multinomial (ψ^{dm})

Fig. 2. The generation process for all posts.

where $n_{t,-j}$ and $\tilde{n}_t$ are the number of times that tag t is generated from perspectives $(X_t = 0)$ and topics $(X_t = 1)$, respectively; $M_{u,-j}^l$ is the number of times that perspective l is adopted by user u; $M_{l,-j}^t$ is the number of times tag t is generated from perspective l; $-j$ indicates taking no account of the current position j.

After enough sampling iterations to burn in the Markov chain, we can estimate the eight parameters in our model: (1) the content-topic distribution θ^d, (2) the topic-word distribution ϕ, (3) the background-word distribution, (4) the binomial distribution π, (5) the topic-dependent word alignment table between a word and a hashtag B, (6) the user-perspective distribution θ^u, (7) the perspective-tag distribution ψ and (8) the binomial distribution λ for any single sample using the following equations:

$$\theta^{(d)} = \frac{M_{k,-d} + \alpha^d}{M_{.,-d} + K\alpha^d}, \quad \phi_{wk} = \frac{M_k^w + \beta}{M_k^{(.)} + W\beta}$$

$$\phi_{wb} = \frac{M_b^w + \beta}{M_b^{(.)} + W\beta}, \quad \pi = \frac{M_{1,-i} + \delta}{M_{.,-i} + 2\delta} \tag{4}$$

$$B_{z,w,t} = \frac{N_{t,w}^z}{\sum_{z'} N_{t,w}^{z'}}, \qquad \theta^{(u)} = \frac{M_{u,-j}^l + \alpha^u}{M_{u,-j}^{(\cdot)} + L\alpha^u}$$

$$\psi_{tl} = \frac{M_{l,-j}^t + \eta}{M_{l,-j}^{(\cdot)} + T\eta}, \qquad \lambda_t = \frac{\tilde{n}_{t,-j} + \gamma}{n_t + \tilde{n}_{t,-j} + 2\gamma},$$

where $N_{t,w}^z$ is the number of occurrences that w is translated to t given topic t.

3.4   Personalized Hashtag Recommendation

In our model, we perform personalized hashtag recommendation as the followings. Given a tweet d with its content $w = \{w_n\}_{n=1}^N$ consisting of N words, paired with its user u. We first perform Gibbs Sampling to iteratively estimate the topic distribution of d (i.e., $\theta^{(d)}$) according to tweet content w. Afterwards, we can rank the hashtags for this post by computing the scores:

$$p(t_m \mid w, u) = p(\lambda_{t_m}) \sum_{c=1}^K \sum_{n=1}^N p(t_m \mid c_m, w, B) \cdot p(c_m \mid \theta_w^{(d)}) \cdot p(w_n \mid w)$$

$$+ [1 - p(\lambda_{t_m})] \sum_{l=1}^L p(t_m \mid p_l) \cdot p(p_l \mid u), \tag{5}$$

where $p(w_n \mid w)$ is the weight of the word w_n in post content w, which can be estimated by *TFIDF* or *IDF*, we apply *IDF* to compute this score. According to the ranking scores, we can suggest the top-ranked personalized hashtags for this post-user either.

4   Experiments and Results

4.1   Datasets

The dataset used for experiment is from a microblogging dataset we collected from Sina-Weibo[3], a popular Twitter-like microblogging system in China. The original dataset contains 20 million microblogs posted by 60,000 users from Sep. 2009 to May. 2013. For prepossessing, we used ICTCLAS2009[4] to segment these microblogs and remove non-standard words (i.e. punctuation marks, urls, at-mentions, etc.) and stop words. Microblogs that do not contain hashtags or containing less than 3 words are removed from the dataset. Since we want to recommend personalized hashtags for microblogs, we filtered out users who posted microblog with hashtags less than 10 times. The remaining dataset contains 112,084 microblogs posted by 6,661 unique users and is used as our final dataset for training and evaluation. Some detailed statistics is shown in Table 1. We divided them into a training set of 101,644 tweets posted by all users in our dataset and a test set of 10,440 tweets. The hashtags actually annotated by users serve as the ground truth.

---

[3] http://weibo.com/.
[4] http://ictclas.org/Down_OpenSrc.asp.

Table 1. Statistics of the dataset used in this paper

#Unique users	6,661
#Microblogs containing hashtags	124,707
#Vocabulary of words	10,7376
#Vocabulary of tags	33,777
#Average number of words in each microblog	27.23
#Average number of words in each hashtag	1.03

4.2 Evaluation Criteria and Experimental Setup

We use Precision (P), Recall (R), and F-value (F) to evaluate the performance of hashtag recommendation methods. These metrics are computed as:

$$Precision = \frac{\#tags \; truly \; assigned}{tags \; assigned \; by \; system} \tag{6}$$

$$Recall = \frac{\#tags \; truly \; assigned}{tags \; manually \; assigned} \tag{7}$$

$$F = \frac{2P \cdot R}{P + R} \tag{8}$$

We ran our model with 500 iterations of Gibbs sampling. After trying a few different numbers of topics and user perspectives while one of them is fixed, we empirically set the number of topics to 10 and the number of perspectives to 80. We use $\alpha^d = 0.1, \alpha^u = 0.1, \delta = 0.1, \psi = 50/L$, and $\beta = 50/K$ as Griffiths and Steyvers [9] suggested. Parameter γ is set to 0.5. As two of our select baselines are variations of topic model, both of them are carefully tuned on the training data also.

4.3 Methods for Comparison

We compare our personal topic translation model (Fig. 1) with 4 baselines and one variation of our model which are described as follows:

- **Naive Bayes (NB)**: This is a representative classification-based method. [7]. We applied this method to model the posterior probability of each hashtag given a tweet.
- **TSTM model**: This is a topical word alignment model proposed by Ding et al. [6], which assumes one tweet have multiple topics.
- **TTM model**: TTM model is a topical translation model described by Ding et al. [5]. We find it is kind of a variation model of TSTM model. To the best of our knowledge, it is the state of art method for hashtags recommendation for microblogs.
- **CF method**: This is a method based on collaborative filtering used by Kywe et al. [10], which combines hashtags of similar users and similar tweets to propose a more personalized set of tags.

– **PTTM_1**: PTTM_1 is a variation of our model, in which we consider one tweet have multiple topics and all the words in tweets are topic-related.

4.4 Experiment Results

Overall Performance. In this subsection, We compared our personal topic translation model (Fig. 1) with those baselines mentioned above. In Fig. 3 we show the precision-recall curves of NB, TSTM, TTM, CF, PTTM_1 and PTTM on the dataset. Each point of a precision-recall curve represents suggesting different number of hashtags M^K, respectively ($M^k = \{1, 2, 3, 5, 10\}$).

Figure 3 clearly shows that PTTM outperforms all the other baseline methods. This indicates the effectiveness of our approach. On one hand, PTTM outperforms the state of art method TTM which implies that it is essential to take user's preference into consideration in the choice of suggesting hashtags for microblogs. On the other hand, PTTM, TTM and PTTM_1 outperform the CF method, it indicates that traditional collaborative filtering method is not suitable for personalized hashtag recommendation problem because of the the shortness of microblog and the diversity of microblogs topics as we concerned. There is an interesting phenomena, when M^k is getting smaller, the advantages of PTTM are more obvious compared to baselines. This implies that when a system is asked to suggest less hashtags for microblog, it is becoming important to take user's preference into account (Table 2).

An additional observation is that TTM outperforms PTTM_1, which may imply that compared to consider user's preference, it is more important to assure that the hashtag is related to the tweet topic. It may illustrate that even though there exits a lot of hashtags are generated from user's perspective, a larger part of hashtags are generated from tweet themes. The last observation is that PTTM outperform PTTM_1 significantly just as TTM outperforms TSTM discriminately, it validates the observation that each microblog tends to cover only one topic.

To further demonstrate the performance of PTTM and other baseline methods, in Table 3, we show the Precision, Recall and F-measure of those models suggesting top-1 hashtag, because the number 1 is near the average number

Table 2. Comparison results of NB, TSTM, TTM, CF, PTTM_1 and PTTM, when suggesting top-1 hashtag.

Method	Precision	Recall	F-measure
NB	0.236	0.231	0.233
TSTM	0.309	0.304	0.306
TTM	0.416	0.411	0.413
CF	0.276	0.267	0.271
PTTM_1	0.341	0.336	0.338
PTTM	**0.447**	**0.441**	**0.443**

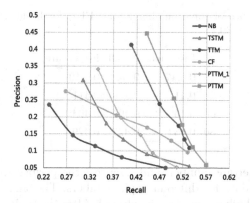

Fig. 3. Precision-recall curves for hashtags recommendation.

of hashtags in dataset. We find that the F-measure of PTTM comes to 0.443, outperforming the state of art method 7.26 % relatively. Since the hashtags are very sparse, we owe the high performance of top 1 recommended hashtag to the incorporation of topic model, as it can suggest hashtags based on the underlying topics of the microblogs.

From Table 3 we observe that: (1) TTM can suggest hashtags that are closely related to the topic, such as "FrenchOpen13" and "Tennis". However, due to not considering user's preference, TTM recommends nearly the same set of hashtags for different users. (2) Taking advantage of considering user's preference into account, PTTM can suggest representative and related hashtags and at the same time guarantee to suite user's taste, such as "Federer Allez". We can see that PTTM can attain a good accuracy, and more, achieves the goal aiming at suggesting hashtags, which would suit both the content and user's preference, for microblogs.

Table 3. Examples of top-5 hashtags suggested by TTM, PTTM.

(User, Topic)	Top-5 hashtags
(2786, Frech Open)	**PTTM**: FrenchOpen13(*), Longines for FrenchOpen daily guess, Federer(+), 1/4FrenchOpen(+), Roger vs Tsonga(+)
	TTM: FrenchOpen13(*), Federer(+), Tennis(+), Longines for FrenchOpen daily guess, 1/4FrenchOpen(+)
(4556, Frech Open)	**PTTM**: Federer Allez(*), FrenchOpen13(+), Longines for FrenchOpen daily guess, Tsonga(+), Roger vs Tsonga(+)
	TTM: FrenchOpen13(+), Federer(+), Longines for FrenchOpen daily guess, Tennis(+), Tsonga(+)

Parameter Influences. There are two crucial parameters in PTTM, the number of topics T and the number of perspectives L. We first fix the number of

Table 4. The influence of topic number T of PTTM for hashtags suggestion when $M^k = 1$ and when $L = 80$.

T	Precision	Recall	F-measure
10	**0.447**	**0.441**	**0.443**
20	0.415	0.409	0.411
60	0.408	0.402	0.404
80	0.433	0.427	0.429

perspectives to a certain number, and then test the performance of the trained model on the test data for different topic numbers. The smallest topic number which leads to the highest accuracy is selected. After the topic is chosen, the perspective number is selected similarly. We demonstrate the performance of PTTM for hashtag recommendation when parameters change in the Tables 4 and 5.

From Table 4, we can see that as the number of topics T varies from $T = 10$ to $T = 80$ when L is fixed to 80, the performance of hahstag suggestion roughly decreases. This shows that the granularity of topics will influence the recommendation performance. When $T = 10$, the performance achieves best, which properly due to the fact that this topic number well covers the topics of the microblogs in the microblogging websites where our corpus crawled from. Hence we set $T = 10$ for our model.

As shown in Table 5, the topic number is fixed to 10, when the perspective number is set with $L = 20$, $L = 60$ or $L = 80$, PTTM achieves the relatively best performance. When $L = 10$, the performance is much poorer. This reveals that compared to the tweet topics, user's perspectives are more diversified. Therefore we set $L = 80$ for our model.

Table 5. The influence of perspective number L of PTTM for hashtags suggestion when $M^k = 1$ and when topic number $K = 10$.

L	Precision	Recall	F-measure
10	0.419	0.413	0.415
20	0.443	0.437	0.439
60	0.445	0.438	0.440
80	**0.447**	**0.441**	**0.443**

5   Conclusions

In this paper, we studied the problem of recommending hashtags for microblogs. Since most of existing work on this task does not take users' preference into consideration, we introduced a novel personal topic translation model which considers the impact of both content and users' preference on hashtag generation.

We compared our model with a classification-based method, a topical translation method, the state-of-the-art models and a traditional collaborative filtering method as well as one variations of our model on a real microblogging dataset. Quantitative evaluations showed that our model could more accurately suggest hashtags for tweet. We also used some case studies to illustrate the effectiveness of the topic factor and the user factor of our model.

References

1. Bandyopadhyay, A., Ghosh, K., Majumder, P., Mitra, M.: Query expansion for microblog retrieval. Int. J. Web Sci. **1**(4), 368–380 (2012)
2. Cui, A., Zhang, M., Liu, Y., Ma, S., Zhang, K.: Discover breaking events with popular hashtags in twitter. In: Proceedings of the 21st ACM International Conference on Information and Knowledge Management, pp. 1794–1798. ACM (2012)
3. Davidov, D., Tsur, O., Rappoport, A.: Enhanced sentiment learning using twitter hashtags and smileys. In: Proceedings of the 23rd International Conference on Computational Linguistics: Posters, pp. 241–249. Association for Computational Linguistics (2010)
4. Diao, Q., Jiang, J., Zhu, F., Lim, E.P.: Finding bursty topics from microblogs. In: Proceedings of the 50th Annual Meeting of the Association for Computational Linguistics: Long Papers, vol. 1, pp. 536–544. Association for Computational Linguistics (2012)
5. Ding, Z., Qiu, X., Zhang, Q., Huang, X.: Learning topical translation model for microblog hashtag suggestion. In: Proceedings of the Twenty-Third International Joint Conference on Artificial Intelligence, pp. 2078–2084. AAAI Press (2013)
6. Ding, Z., Zhang, Q., Huang, X.: Automatic hashtag recommendation for microblogs using topic-specific translation model. Proc. COLING **2012**, 265–274 (2012)
7. Garg, N., Weber, I.: Personalized, interactive tag recommendation for flickr. In: Proceedings of the 2008 ACM Conference on Recommender Systems, pp. 67–74. ACM (2008)
8. Godin, F., Slavkovikj, V., De Neve, W., Schrauwen, B., Van de Walle, R.: Using topic models for twitter hashtag recommendation. In: Proceedings of the 22nd International Conference on World Wide Web Companion, pp. 593–596. International World Wide Web Conferences Steering Committee (2013)
9. Griffiths, T.L., Steyvers, M.: Finding scientific topics. Proc. Nat. Acad. Sci. USA **101**(Suppl 1), 5228–5235 (2004)
10. Kywe, S.M., Hoang, T.-A., Lim, E.-P., Zhu, F.: On recommending hashtags in twitter networks. In: Aberer, K., Flache, A., Jager, W., Liu, L., Tang, J., Guéret, C. (eds.) SocInfo 2012. LNCS, vol. 7710, pp. 337–350. Springer, Heidelberg (2012)
11. Li, T., Wu, Y., Zhang, Y.: Twitter hash tag prediction algorithm. In: ICOMP11 - The 2011 International Conference on Internet Computing (2011)
12. Liu, Z., Liang, C., Sun, M.: Topical word trigger model for keyphrase extraction. In: COLING, pp. 1715–1730 (2012)
13. Mazzia, A., Juett, J.: Suggesting hashtags on twitter (2009)
14. Tariq, A., Karim, A., Gomez, F., Foroosh, H.: Exploiting topical perceptions over multi-lingual text for hashtag suggestion on twitter. In: The Twenty-Sixth International FLAIRS Conference (2013)

15. Wang, X., Wei, F., Liu, X., Zhou, M., Zhang, M.: Topic sentiment analysis in twitter: a graph-based hashtag sentiment classification approach. In: Proceedings of the 20th ACM International Conference on Information and Knowledge Management, pp. 1031–1040. ACM (2011)
16. Zangerle, E., Gassler, W., Specht, G.: Recommending#-tags in twitter. In: Proceedings of the Workshop on Semantic Adaptive Social Web (SASWeb 2011), CEUR Workshop Proceedings, vol. 730, pp. 67–78 (2011)
17. Zangerle, E., Gassler, W., Specht, G.: On the impact of text similarity functions on hashtag recommendations in microblogging environments. Soc. Netw. Anal. Min. **3**(4), 1–10 (2013)
18. Zhao, W.X., Jiang, J., Weng, J., He, J., Lim, E.-P., Yan, H., Li, X.: Comparing twitter and traditional media using topic models. In: Clough, P., Foley, C., Gurrin, C., Jones, G.J.F., Kraaij, W., Lee, H., Mudoch, V. (eds.) ECIR 2011. LNCS, vol. 6611, pp. 338–349. Springer, Heidelberg (2011)

Hybrid Model Based Influenza Detection with Sentiment Analysis from Social Networks

Xiao Sun[1]([⊠]), Jiaqi Ye[2], and Fuji Ren[3]

[1] School of Computer and Information, Hefei University of Technology, Hefei, China
sunx@hfut.edu.cn
[2] Hefei University of Technology and Tokushima University, Tokushima, Japan
[3] School of Computer and Information, School of Management,
Hefei University of Technology, Hefei 230009, Anhui, China
ren@is.tokushima-u.ac.jp

Abstract. Sina microblog is a popular microblogging service in China, which could provide perfect reference sources for flu detection due to its' real-time characteristic and large number of active users posting about their daily life continually. In this paper, we investigate the real-time flu detection problem and propose a flu detection model with emotion factors(sentiment analysis) and sematic information (Em-Flu model). First, we extract flu-related microblog posts automatically in real-time using a trained SVM filter. For posts classification, we also adopt association rule mining to extract strongly associated features as additional features of posts to overcome the limitation of 140 words, including sentiment analysis information which can help to classify the posts without explicit flu-related features. Then Conditional Random Field model is revised and applied to detect the transition time of flu that we can find out which place is more likely for influenza outbreak and when is more likely for influenza outbreak in one city or a province in China. Experimental results on detecting flu situation during certain time in some locations show the robustness and effectiveness of the proposed model.

Keywords: Influenza detection · Conditional random field · Transition time detection · Social web mining · Public health

1 Introduction

Influenza is a highly contagious acute respiratory disease caused by influenza virus. As the highly genetic variation, influenza can cause global epidemic, which not only brought huge disasters to people's life and health, but also have significant disruptions for country economy. There are about 10–15 % of population who get influenza every year, which results in up to 50 million illnesses and 500,000 deaths in the world each year. Influenza is a worldwide public health problem and there are no effective measures to control its epidemic at present. The prevalence of influenza in China is one of the most notable problems.

© Springer Science+Business Media Singapore 2015
X. Zhang et al. (Eds.): SMP 2015, CCIS 568, pp. 51–62, 2015.
DOI: 10.1007/978-981-10-0080-5_5

Nowadays influenza surveillance systems have been established via the European Influenza Surveillance Scheme (EISS) in Europe and the Centre for Disease Control (CDC) in the US to collect data from clinical diagnoses. The research of forecasting methods started relatively late in China via Chinese National Influenza Center (CNIC) and all these systems have approximately two week's delay. The need for efficient sources of data for forecasting have increased due to the Public health authorities' need to forecast at the earliest time to ensure effective treatment. Another popular surveillance system is Google's flu trends service [14] which is web-based click flu reporting system. Google's flu trend adopts keywords statistic and uses linear model to link the influenza-like illness visits. Recent works have demonstrated that prediction of varieties of phenomena can be made by using social media data. Based on the real-time data of microblog, there has been some applications such as earthquake detection [1], public health tracking [2,3] and also flu detection [5,8]. The measures of collecting clinical diagnoses and web-based clicks on key word with linear model are quite good but not fair enough. Some extra knowledge in post text such as semantic information in post is neglected, which might also be an important indication for flu detection. Our research tries to use the big real-time social network data as resources and design a discriminative machine learning model with emotional factors and semantic information, which could improve the effectiveness of finding the break point of influenza.

This work is distinguished by two key contributions. The first is the combination of supervised an unsupervised for extracting samples and training dataset. The unsupervised sample extraction method help to rectify manual labeled factors and the supervised method take the advantage of preliminary stage with human priori knowledge avoiding the random mistakes. The second contribution is bring in CRF-label bias and Markov length bias to help find global optimum.

2 Em-Flu Model

Existing works on flu prediction pay more attention on how to fit real world CDC or ILI data, but seldom consider the algorithm for breakout detection. Spatial information is seldom considered because influenza is an acute infectious disease and sematic or emotion factors in text are also out of considerations thus lots of useful features would be ignored. In order to overcome this problem, in this paper, we introduce a supervised discriminative approach called Em-flu model for early stage flu detection.

Come down to influenza infection process, CDC mentions "Most healthy adults may be able to infect other people beginning 1 day before symptoms develop and up to 5 to 7 days after becoming sick. Children may pass the virus for longer than 7 days. Symptoms start 1 to 4 days after the virus enters the body."[1] Martinez [16] talked about a simple two-stage model in his work. We modeled this process information in a four-phase switching model, which describes the corresponding stages of flu, i.e. non-epidemic phase (NP), rising

---

[1] http://www.cdc.gov/flu/about/disease/spread.htm.

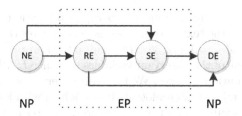

Fig. 1. Four-phase switch pattern of flu

epidemic phase (RP), stationary epidemic phase (SP) and declining epidemic phase (DP). Apparently we focus on the switch point from non-epidemic phase (NP) to epidemic phase (EP). Generally speaking, the course of influenza may last a week or two, for a single microblog user, it is supposed that his(or her) microblog contents will record a series of flu state and emotional feelings when user is sick or catching flu. When a person gets the flu, he will usually go through four flu phrases: NE, RE, SE, DE; despite of some keywords, the underlying emotion embedded in microblog test of these four flu phases would naturally change from on phase to another. All these individuals' data could be combined into dataset organized by time. In Fig. 1 we describe the switch pattern of our assumption. Normally one person is at the NP stage. Once infected flu the symptom begin to start (RP) and at the same time or later time people around infected person could be newly infected. Then a few days later the first infected person come into (SP) that means the symptoms are almost out and keeps to the next stage. After a week more or less the symptoms are declining (DP) then come to the NP stage.

Our approach assumes microblog users as "sensors" and collective posts containing flu keywords as early indicators. In some way, as real data from a mass of real users are adopted for flu detection, This make our approach more like a crowd sourcing way to get useful data. Further more, Conditional Random Fields (CRFs) is revised and adopted for detecting the transition time of flu. As the output of CRFs model is modeled on the whole input sequence, so that the proposed algorithm can capture flu outbreaks more promptly and accurately compared with markov or other baseline models. Based on the proposed algorithm, a real-time flu detection and visualization system is also built. For early stage flu detection, a probabilistic graphical approach is adopted and the key of the flu detection task is to detect transition time of flu from non-epidemic (NP) phase to epidemic phase (EP) and so on.

Conditional random fields [13] is a type of discriminative undirected probabilistic graphical model. It is used to encode known relationships between observations and construct consistent interpretations. It is often used for labeling or parsing of sequential data, such as natural language text or biological sequences and in computer vision.

Basically, our model is based on a segmentation of the series of differences into an epidemic and a non-epidemic phase using CRF model. Suppose we collected flu related microblog text data from N geographical location. For each location $i \in [1, N]$, we segment the text into time series. $Z_{i,t}$ denotes the current flu phase of location at time t. $Z_{i,t} = 0, 1, 2, 3$ correspond to the four phases of flu: NE, RE, SE and DE. $E_{i,t}$ is one of the observant variables sequence E, which represent the daily emotion distribution changes. We take this observation variable to denotes the rising risk of flu related microblog at time t, joined with location i. We hope this emotion changes will manifest better experimental result than only consider flu-related posts number changes.

Let E_{-1}, E_1, E_1 respectively denote the daily post numbers of positive, neural and negative emotion posts at the place t time i. Then we evaluate the flu emotion level by the following function:

$$E_{i,t} = \log(E_{+1} - E_{-1})/(E_{-1} + E_0 + E_{-1}) \tag{1}$$

$$\Delta E_{i,t} = (E_{i,t} - E_{i,t-1})/E_{i,t-1} \tag{2}$$

We consider the emotion effect which describes the people's emotion state when they post flu related microblogs on each day within the week. We focus on the characteristics for the dynamics of different flu phases. Let $k = day(t)$ denote the day within the week. For time t, the observations $\Delta E_{i,t}$ are modeled as a Gaussian distribution with mean L_k and variance δ_k^2. Based on Bayesian perspective, we give a Gaussian prior for L_k and an $Inverse$-χ^2 prior for δ_k^2 based on collected data. W_k denotes the mean of $\Delta E_{i,t}$ with $day(t) = k$ within that time period and G_k^2 means the variance based on previous data.

$$\Delta E_{i,t} \sim N(L_k, \delta_k^2) \tag{3}$$

$$L_k \sim N(W_k, \phi_k^2) \tag{4}$$

$$\delta_k^2 \sim Inv - \chi^2(v_k, G_k^2) \tag{5}$$

The underlying idea of Conditional Random Field switching models is to associate each $E_{i,t}$ with a random variable $Z_{i,t}$ that determines the conditional distribution of $Z_{i,t}$ given the whole observe sequence E. In our case, each $Z_{i,t}$ is an unobserved random variable that indicates which flu phase the $i - th$ location is now(at time t) in. Moreover, the unobserved sequence of $Z_{i,t}$ follows a four-stage Markov chain with transition probabilities. For location i, $N(i)$ denotes the subset containing its geographical neighbors. We simplify the model by only considering bordering states in $N(i)$.

$$\Delta N_{i,t} = \frac{1}{n} \sum \Delta E_{k,t-1}(k \in N(i)) \tag{6}$$

Then formula 6 means weight of the provinces around location i which provide the risk of influenza infections. $\Delta N_{i,t}$ is also an observant variable, which could be easily obtained from original corpus by statistic methods.

The spatial information of social network is modeled in a unified CRFs Markov Network in Fig. 2, where the flu phase for location i at each time is not only depending upon its previous phase, but also on its location neighbors.

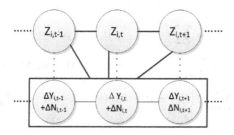

Fig. 2. Graphical Illustration for flu detection

In this work, for simplification, we only treat bordering states as neighbors. Since the influence from non-bordering locations can also be transmitted through bordering ones, we take the sum of the variable $\Delta N_{i,t}$ and $\Delta E_{k,t-1}$ as the final influence state in location i. such simplification make sense and experimental results also demonstrate this point. For location i at time t, the probability that $Z_{i,t}$ takes on value Z is illustrated to integrate the spatial information in a unified framework:

$$P(Z_{i,t}|\Delta N_{i,t} + \Delta E_{i,t}, \lambda) = \propto \exp(\sum_j \lambda_j u_j(Z_{i-1}, Z_i, \Delta N_{i,t} + \Delta E_{i,t}, i)$$
$$+ \sum_k \mu_k v_k(Z_i, \Delta N_{i,t} + \Delta E_{i,t}, i)) \tag{7}$$

$u_j(Z_{i-1}, Z_i, \Delta N_{i,t} + \Delta E_{i,t}, i)$ denotes the transfer feature function of observant tag sequences between -1 to 1. $v_k(Z_i, \Delta N_{i,t} + \Delta E_{i,t}, i)$ denotes the condition feature function of observant tag sequences at location i. These two functions can be combined into a unified expression $f_j(Z_{i-1}, Z_i, \Delta N_{i,t} + \Delta E_{i,t}, i)$ which could act as the feature function for CRFs model.

$$P(Z_{i,t}|\Delta N_{i,t} + \Delta E_{i,t}, \lambda) = \frac{1}{M(\Delta N_{i,t} + \Delta E_{i,t})}$$
$$\exp\left[\sum_{i=1}^n \sum_j \lambda_j f_j(Z_{i-1}, Z_i, \Delta N_{i,t} + \Delta E_{i,t}, i)\right] \tag{8}$$

$$M(\Delta N_{i,t} + \Delta E_{i,t}) = \sum_j \exp\left[\sum_{i=1}^n \sum_j \lambda_j f_j(Z_{i-1}, Z_i, \Delta N_{i,t} + \Delta E_{i,t}, i)\right] \tag{9}$$

We inference the parameters along standard evaluation method for example sampling L_k and δ_k^2:

$$[L_k|-] = [L_k] \cdot \prod_{\substack{Z_{i,t}=0,2 \\ day(t)=k}} [\Delta E_{i,t}|\delta_k^2, L_k] \propto N\left(\frac{\frac{M_k}{\delta_k^2} + \frac{M_k}{\phi_k^2}}{\frac{N_k}{\delta_k^2} + \frac{1_k}{\phi_k^2}}, \frac{1}{\frac{N_k}{\delta_k^2} + \frac{1_k}{\phi_k^2}}\right) \tag{10}$$

$$[\delta_k^2|-] = [\delta_k^2] \cdot \prod_{\substack{Z_{i,t}=0,2 \\ day(t)=k}} [\Delta E_{i,t}|\delta_k^2, L_k] \propto Inv\text{-}\chi^2\left(\mu_k + N_k, \frac{\mu_0\delta_k^2 + V_k}{\mu_k + N_k}\right) \tag{11}$$

where

$$N_k = \sum_{i,t} I(Z_{i,t} = 0, 2; day(t) = k) \qquad (12)$$

$$W_k = \sum_{i,t} I(Z_{i,t} = 0, 2; day(t) = k)\Delta E_{i,t} \qquad (13)$$

$$V_k = \sum_{i,t} I(Z_{i,t} = 0, 2; day(t) = k)(\Delta E_{i,t} - L_k)^2. \qquad (14)$$

3   Data Preparation

We extend our earlier work on Sina microblog data acquisition method and developed a dedicated crawler to fetch data at regular time intervals [15]. We fetched microblog records containing indicator words shown in Table 1 and collect about 4 million flu-related microblog starting from January 2013 to January 2014. Location details can be obtained from the profile page. We select microblogs whose location are in China and discard those ones with meaningless locations.

Table 1. Indicator seed words set for data collection

Indicator words	
止咳药(pectoral)	抗生素(antibiotic)
输液(transfusion)	喉咙疼(sore throat)
伤风(cold)	感冒(influenza)
流涕(running nose)	发烧(fever)
流感(flu)	发高烧(high fever)
咳嗽(cough)	鼻涕(snot)

Not all microblog containing indicator keywords indicate that the user is infected. Meanwhile the indicator words list may not be perfect, so the indicator words list needs to expand from the data we have and the dataset needs to be processed before be used for our task.

The words in Table 1 will be used as seed words to find the initial dataset and then computing vector in the dataset to find other keyword which can be the representations of seed words. We take Google word2vec model tools [12] for these seeds keyword extension. After iterating and training in corpus of 20 million words obtained and then clustering the keyword for the need of computing the distance of different words we obtained with the most relevant 100 words.

Step1: Preprocess: word segmentation, remove the stop words, and combine the text to a document as input for next step.

Step2: Word clustering: take word2vec to deal with input data. The parameters set are as follows: (-classed: 50), (-cbow: 0), (-negative: 0), (-hs: 1), (-sample: l), (-window: 5), (-size: 100). The word cluster results in vectors:

$$ClusterResult = (< word_1, C_1 >< word_2, C_2 > ... < word_n, C_n >) \qquad (15)$$

Where n means the words sum total, C_i means the class which $word_i$ belongs to. After iterating and training in corpus of 20 million words obtained and then

Table 2. Indicator seed words set for data collection

发烧(fever)	发高烧,低烧,拉肚子...High fever, low fever, diarrhea
输液(transfusion)	打针,吊瓶,住院...Take injection, drip, in hospital
伤风(cold)	风寒,头疼,流行性...Wind chill, headache, epidemic
流涕(running nose)	鼻涕,头疼,纸巾... Snot, headache, facial tissue
流感(flu)	禽流感,H1N1,病毒...Bird flu, H1N1,virus
咳嗽(cough)	干咳,干呕,痰...Dry cough, retching, phlegm

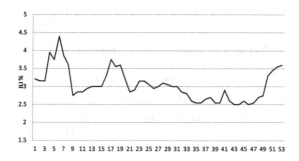

Fig. 3. CNIC influenza weekly data of the year 2013

clustering the keyword for the need of computing the distance of different words we obtained with the most relevant 100 words in Table 2.

In this way, not only words list could be expanded and adapt the changes of cyber word but also help recall more influenza related corpus which would improve the degree of accuracy in our system. The necessity of filtering in real-time task has been demonstrated in many existing works [1,6]. To filter out these bias microblog content, we first prepared manually labeled training data, which was comprised of 3000 microblog records containing key words. Three annotators are responsible for assigning positive or negative label to every post in training dataset and test dataset. One post is labeled as a positive only when it meets the requirements.

In China, the influenza surveillance department is Chinese National Influenza Center. Every week CNIC will release Chinese influenza weekly report. We collect the reports data among the year of 2013 and analysis the official reports to clear up the whole year influenza infection changes. In Fig. 3, we could see the influenza weekly data of the year 2013.

We know that there is mass noise in the crowded microblog contents. We built a classifier based on support vector machine which is a well-known supervised learning model with associated learning algorithms that analyze data and recognize patterns, used for classification and regression analysis. Furthermore we manual calibrate data according to the CNIC data that we collected because the CNIC data relies on the true medical surveillance system. We use a valid tool named libSVM with a linear kernel to handle this pre-filter mission. In terms of

Table 3. Result of different combinations of features for filtering

Features	Accuracy	Precision	Recall
A	84.21 %	82.31 %	89.40 %
B	85.10 %	84.92 %	87.00 %
A+B	87.40 %	88.75 %	89.64 %
A+B+C	84.20 %	87.56 %	89.64 %
A+B+D	76.59 %	88.43 %	87.33 %
A+B+C+D	72.05 %	73.37 %	88.17 %

Table 4. Results on BN, HMM and SVM

Models	Accuracy	Precision	Recall
BN	80.12 %	83.23 %	87.22 %
HMM	83.85 %	87.29 %	80.94 %
SVM	87.40 %	88.75 %	89.64 %

SVM, the training data set D with points is defined as below:

$$D = \{(x_i, y_i)|x_i \in R^p, y_i \in \{+1, -1\}\}_{i=0}^n \qquad (16)$$

Where x_i is a P-dimensional real vector and y_i is the label of the point x_i, indicating to which class belongs. And the classification function of SVM is:

$$f(x) = sign(\sum_i a_i y_i x_i^T + b) \qquad (17)$$

We employ the following simple text-based features Feature A: Collocation features, representing words of query word within a window size of two. Feature B: unigrams, denoting presence or absence of the terms from dataset. Feature C: position of keywords. Feature D: microblog length count. Each of these three annotators individually labeled a post x as negative (-1) or positive $(+1)$ influenza like post described as A, B, C and D. Each post was given the final label by the following function, where the positive value of L indicates a positive influenza post, while the negative value of L indicates a negative influenza post. Where the value of $f(x)$ indicates the point's class, a_i is Lagrange multiplier and b is the intercept. Performances for different combinations of features are illustrated at Table 3. We observe that A+B is much better than other features' combination. So in our following experiments, microblog are selected according to a classifier based on feather A+B. We also take BN and HMM model for comparison with the SVM model which taken the best feature selection. The results are shown in Table 4.

Based on the techniques mentioned above, we can filter the influenza-like microblogs easily. For these flu-related microblog records, we generate another microblog web crawler to deal with every record. For every record's user, we use

Table 5. The result of training and open test on phase detection

Phases	P	R	F
NE	78.10 %	67.23 %	72.26 %
RE	81.54 %	71.34 %	76.10 %
SE	67.97 %	72.78 %	70.29 %
DE	75.23 %	70.22 %	72.64 %

this tool to backup user's microblog content and cut records by a window of time with one week before and after the flu-related microblog record which we had captured. We collected 1000 user's microblog content over all more than 30 thousand records as four phases switching identify experimental material later.

4 Experiments and Data Analysis

The main goal of our task is to help raise an alarm at those moments when there is a high probability that the flu breaks out. In real time situations, for each time, available data only comes from the previous days, and there is no known information about what will happen in the following days or week. By adding the data corpus day by day, we calculate the posterior probability for transiting to epidemic states based on previous observed data. To achieve our target, we design two train projects for two CRF model. One is to detect which phase is the coming microblog content in and the other model is to predict whether there is an outbreak of influenza.

First, we take the 30 thousand records we have collected through preprocessing, such as word segmentation, filtering stop words. Then we manually label the words list by the NE, RE, SE, DE and N. O denotes other situation and unknown judgment. The judging criteria are not only obeying the Gaussian process which is mentioned in Sect. 2, the emotion factors need to be considered as well. We have a table of emotional words with emotion numerical value to help assist the label process. Thus there are three rows in the data corpus: one is the original text, second row is phase, the third is sentiment tag.

We use two thirds of the labeled records as training data sheet, the rest as testing data. The open test result of first model and is shown in Table 5. From the Table 5 we can find that the stage RE is easier to detect correctly because the precision and F value are higher than other three phases. Thus when we have a new microblog and we can test the microblog to find out whether it is influenza-like post but also which phase it is in. If we find the rapid rising tendency of RE phase posts we can say that the influenza pandemic will come soon.

From the data we tested, Fig. 4 shows the global distribution of DE, SE and RE in the year of 2013. The left hand side figure corresponds to number of flu-related microblog records overtime. Purple symbols denote the phase of RE, red symbols denote the phase of SE and white symbols denote the phase of DE. We can find that week 4, 10, 16 and 47 may be the breakpoint of influenza.

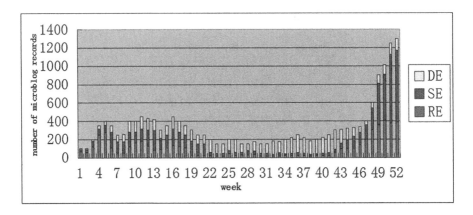

Fig. 4. Predictions of the year 2013 (Color figure online)

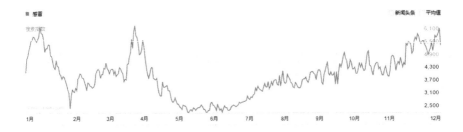

Fig. 5. Searching Result on Baidu Index platform

Compared to the Fig. 3, week 3, 17 and 51 is the rising edge that suggests the flu breakpoint.

Figure 5 shows the result of searching key words like influenza on Baidu Index platform. Like the techniques of Google flu trend service, Baidu Index platform uses Chinese search engine to collect user's key words counting. The fluctuation of Baidu index ranges greatly. It suggests week 2, 15 is the search flu-related key words breakpoint and in week 32 to 52 the searching activity rising steadily. As this is the overall flu trend situation, there is a lot of factors that influence the results. We need more detailed process to find out the real breakpoint of influenza because flu is a infectious disease that from point to surface and it has obvious infection pattern.

The second model we try to design is to predict whether there is an outbreak of influenza and what the chances are. We focus on the phase RE because if we find the phase RE rises rapidly in location i or the neighbors' RE value maintained at a dangerous level, there is confidence that the flu is coming soon on account of the highly infectious.

For comparison, we employ the following baseline in this paper: Average: Uses the average frequency of microblog records containing keywords based on previous years as the threshold. Two-Phase: A simple version of approach but

Table 6. The examples of our model on week 40–52's microblog data from different provinces Province

•	Peak time	Average	Two-phase	Em-flu
Anhui	Nov.25	Nov.15	Nov.10	Nov.08
Zhejiang	Nov.06	Nov.01	Oct.17	Oct.10
Guangdong	Oct.20	Nov.05	Oct.10	Oct.07
Fujian	Nov.10	Oct.18	Oct.19	Sep.20
Jiangsu	Oct.15	Nov.03	Nov.03	Sep.28
Shandong	Nov.06	Oct.20	Oct.27	Oct.05
Liaoning	Oct.22	Oct.25	Oct.07	Oct.09
Shanghai	Nov.17	Nov.15	Oct.25	Sep.24
Jiangxi	Nov.04	Oct.23	Nov.01	Sep.18
Henan	Nov.18	Nov.19	Oct.27	Oct.07

using a simple two-phase in Markov network. In Table 6 we present the performance from different baselines along with our algorithm based on 10 provinces microblog data in China. Peak time denotes the moment when the frequency containing keywords reaches the top. From Table 6, we can identify that for most provinces our model would detect the breakout of influenza earlier than other method. This validates our assumption that by considering neighboring influence can achieve a better flu prediction performance.

5   Conclusion

In this paper, our interest has been to introduce a supervised CRF model based on four phases, and microblog emotional factors are appended in the model to help detect early stage flu epidemics on Sina Microblog. We test our model on real time datasets for multiple applications and experiments results demonstrate the effectiveness of our method. We now comment on possible extensions to this study. The first possibility could be to explore whether the probability of being in an epidemic phase could depend on the rate of the previous week. Another potential extension of our model would include a multivariate or a spatial component that could help us to explore any geographical disaggregation of the rates. Futhermore, we will try to adopt some neural network model for sequence labeling, which might perform better than CRF.

Acknowledgment. The work is supported by National Natural Science Funds for Distinguished Young Scholar(No.61203315), This work was partially supported by JSPS KAKENHI Grant Number 15H01712. This work was supported by the Open Project Program of the National Laboratory of Pattern Recognition (NLPR).

References

1. Sakaki, T., Okazaki, M., Matsuo, Y.: Earthquake shakes Twitter users: real-time event detection by social sensors. In: Proceedings of the 19th international conference on World wide web, pp. 851–860. ACM (2010)
2. Collier, N.: Uncovering text mining: a survey of current work on web-based epidemic intelligence. Global Public Health **7**(7), 731–749 (2012)
3. Paul, M.J., Dredze, M.: You are what you Tweet: Analyzing Twitter for public health. In: ICWSM (2011)
4. Achrekar, H., Gandhe, A., Lazarus, R., et al.: Predicting flu trends using twitter data. In: 2011 IEEE Conference on Computer Communications Workshops (INFOCOM WKSHPS), pp. 702–707. IEEE (2011)
5. Culotta, A.: Towards detecting influenza epidemics by analyzing Twitter messages. In: Proceedings of the first workshop on social media analytics, pp. 115–122. ACM (2010)
6. Aramaki, E., Maskawa, S., Morita, M.: Twitter catches the flu: detecting influenza epidemics using Twitter. In: Proceedings of the Conference on Empirical Methods in Natural Language Processing, pp. 1568–1576. Association for Computational Linguistics (2011)
7. Lamb, A., Paul, M.J., Dredze, M.: Separating fact from fear: tracking flu infections on twitter. In: Proceedings of NAACL-HLT, pp. 789–795 (2013)
8. Achrekar, H.: Online Social Network Flu Tracker a Novel Sensory Approach to Predict Flu Trends. University of Massachusetts, Lowell (2012)
9. Aschwanden, C.: Spatial simulation model for infectious viral diseases with focus on SARS and the common Flu. In: HICSS (2004)
10. http://www.google.org/flutrends/, May 2015
11. Lazer, D., Kennedy, R., King, G., Vespignani, A.: The parable of google flu: traps in big data analysis. Science **343**(6176), 1203–1205 (2014)
12. https://code.google.com/p/word2vec/, May 2015
13. Lafferty, J., McCallum, A., Pereira, F.: Conditional random fields: probabilistic models for segmenting and labeling sequence data. In: Proceeding of 18th International Conference on Machine Learning, pp. 282–289. Morgan Kaufmann (2001)
14. Cook, S., et al.: Assessing Google flu trends performance in the United States during the 2009 influenza virus A (H1N1) pandemic. PloS One **6**(8), e23610 (2011)
15. Sun, X., Ye, J., Tang, C., et al.: The method and application of Sina microblogging big data grabbing based on Mulit-strategy. J. Hefei Univ. Technol. (Natural Science) **17**(10), 1210–1215 (2014)
16. Martinez-Beneito, M.A., et al.: Bayesian Markov switching models for the early detection of influenza epidemics. Stat. Med. **27**(22), 4455–4468 (2008)

Do Photos Help Express Our Feelings: Incorporating Multimodal Features into Microblog Sentiment Analysis

Tianyi Liu[✉], Fei Jiang, Yiqun Liu, Min Zhang, and Shaoping Ma

State Key Laboratory of Intelligent Technology and Systems,
Tsinghua National Laboratory for Information Science and Technology,
Department of Computer Science and Technology, Tsinghua University,
Beijing 100084, China
skywalker_lty@163.com, f91.jiang@gmail.com
{yiqunliu,z-m,msp}@tsinghua.edu.cn

Abstract. Because of the interest in discovering public moods and opinions in both industrial and academic researches, sentiment analysis of microblogs has become one of the major concerns in Web data mining and natural language processing studies. Although a large part of the microblog posts contain non-textual components such as images, emoticons and location information, most existing works rely on textual information only to generate sentiment analysis results. Different from these efforts, we focus on the influence of other sources of information in sentiment analysis, especially the images from social media, which are commonly posted by users along with texts. Having noticed that images reinforce sentiment expression along with text in microblog environment, we propose a unified model to extract the features of text and image together. Learning based approaches are then adopted to finish sentiment analysis tasks such as subjectivity classification. Experimental results based on practical microblog data show that features extracted from images help gain better sentiment analysis results.

Keywords: Microblog Sentiment Analysis · Subjectivity classification · Image Feature Extraction · Text Feature Extraction · Emoticon Space Model

1 Introduction

Nowadays, social media has become popular because of convenience, agility and wide-spreading. People tend to express their ideas through Microblog and Twitter, which are regarded as two kinds of the most popular social media. Report from CNNIC[1] shows that there have been 249 million microblog users in China at the end of 2014. Statistics also indicate that, Sina Weibo, the most popular

---

[1] 35th China Internet Development Statistics Report.

© Springer Science+Business Media Singapore 2015
X. Zhang et al. (Eds.): SMP 2015, CCIS 568, pp. 63–73, 2015.
DOI: 10.1007/978-981-10-0080-5_6

microblog platform in China, own 167 million active users per month. Approximately 100 million microblog posts are generated every day. In a post, users can express their ideas on specific topics, convey their feelings about social events and forward opinions from other users.

Therefore, the content of a microblog post is worth mining and analyzing. Research on microblog contents can assist in performance of predicting housing price or stock market, acquiring publics evaluation for a company and even analyzing publics attitudes towards social events or government policies. As a main research aspect, sentiment analysis is crucial in analyzing contents of microblog posts and thus has become a hot research area these years.

In sentiment analysis, several important issues are studied. For example, analyzing what kind of emotion is expressed in microblog, identifying whether a post is subjective or objective (subjectivity classification), identifying whether a post is positive or negative (polarity classification) and so on.

To solve these problems, Machine Learning methods are widely adopted in sentiment analysis because they are proved effective in solving classification issues especially for binary-classification. Previous researches [2,10] have studied the sentiment of text in microblog posts. However, the image in posts are not widely studied. Existing works that are relevant to image processing mainly focus on affective computing. They utilize color and texture as image features for sentiment analysis.

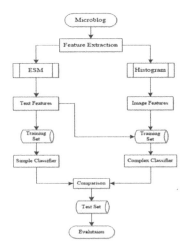

Fig. 1. Main procedure

In this paper, we propose a method to perform sentiment classification. Previous researches have mostly focused on text analysis. However, in microblog environment, images should not be ignored. The images uploaded by users also express emotions along with the text in a microblog post. Therefore, we perform sentiment analysis with text and image features together. In experiment, we have a microblog corpus that comes from Sina Weibo. Each data in corpus is a post. We extract text features by applying Emoticon Space Model (ESM) and extract statistical image features like histogram. The label of each post is annotated manually. By this means, we get a complete microblog dataset whose each posts features are from text and image. For training model, we select SVM to train classifiers. First, we train and acquire a simple classifier with text features only. Then we train and acquire a complex classifier with text features and image features. The experiment results show the performance of classifier with image features is better than classifier without image features, which proves that images also have influences on emotion expressing and polarity classification. In microblog sentiment analysis, we should synthetically consider text and image features to better recognize the sentiment. Figure 1 shows the main procedure of our work.

2 Related Work

Both supervised and unsupervised learning methods have been used for microblog sentiment analysis. However, compared with unsupervised learning, classifiers trained by supervised learning method perform better in microblog environment.

Among supervised learning method, a state-of-the-art method has introduced deep learning into sentiment analysis. [5] proposes an innovative model to capture words association and build up sentiment transfer model to reinforce the capture of text association. Compared with traditional method, the performance of his method is at least the same as previous research. However, his work avoids burdensome manually annotation to some extent.

Text analysis method has been widely studied. Apart from studying on pure text, some researchers [2,10] have synthetically considered pure text and emoticons contained in microblogs. In fact, emoticons include text emoticons like :) and graphical emoticons like 😊. They are both signals for expressing emotion. In [10] 's research, they build up Chinese microblog sentiment corpus by combining sentiment words and graphical emoticons. They train Bayes classifier and make progress by utilizing concepts of entropy. Their work has achieved high precision and recall. However, they only consider the text and emoticons. The limitation is that they havent done further research on images in microblogs.

Image analysis in microblogs is rarely studied. Some existing works on image sentiment focus on affective computing. Colors and textures are studied as sentiment features. In [9], researchers extract low-level visual features of image to construct visual feature space. In [11] 's work, they synthetically consider both text and image features in microblogs. In images' study, they extract the color and texture information. Saturation and brightness are also discussed in their work. They propose a new neighborhood classifier and compare it with traditional classifier such as SVM and NaiveBayes. Results show their classifier get better performance in classification task with text and image features.

In sentiment analysis, there are multimodal features besides text and image. In [8] 's research, visual, acoustic and linguistic features are used in sentiment analysis. They focus on video reviews from website such as Youtube, Facebook and Amazon. They introduced a new multimodal dataset consisting of sentiment annotated utterances extracted from video reviews, where each utterance is associated with a video, acoustic, and linguistic datastream. Our experiments show that sentiment annotation of utterance-level visual datastreams can be effectively performed, and that the use of multiple modalities can lead to error rate reductions of up to 10.5 % as compared to the use of one modality at a time.

And there are some researches focus on videos on Internet. In [6] 's work, they study the constant data flow in videos on websites. They addresses the task of multimodal sentiment analysis, and conducts proof-of-concept experiments that demonstrate that a joint model that integrates visual, audio, and textual features can be effectively used to identify sentiment in Web videos.

Other related works on Twitter sentiment analysis [1,4] are also interesting. Some researchers study opinions in Twitter [3,7]. In our research, considering the real environment in Sina Weibo platform, we mainly discuss the text and image contained in microblogs. Apart from baseline work that dealing with text and emoticons, we also explore statistical method to analyze and extract image information.

3   Text Feature Extraction

To deal with texts and better analyze the signals contained in emoticons, we utilize ESM (Emoticon Space Model), a semi-supervised model firstly proposed by Fei Jiang in [2]. In this model, we select a relatively large number of commonly employed emoticons with and without clear emotional meanings to construct an emoticon space, where each emoticon serves as one dimension. The core of ESM is projection from text to emoticons space. Concretely, texts are projected into the emoticon space according to the semantic similarity between words and emoticons, which can be learned from a massive amount of unlabeled data. Besides, an assumption is made that posts with similar sentiments have similar coordinates in this space.

Our process of text feature extraction is based on ESM. First, both words and emoticons have a distributed representation that provides an effective way to learn the semantic similarity between words and emoticons. Concretely, the cosine similarity is used as the measurement of similarity between the representation vectors, which can be formalized as

$$similarity(\boldsymbol{w_i}, \boldsymbol{e_j}) = \frac{\boldsymbol{w_i} \cdot \boldsymbol{e_j}}{|\boldsymbol{w_i}| \, |\boldsymbol{e_j}|} \qquad (1)$$

w_i and e_j are the representation vectors of word i and emoticon j. We use this semantic similarity as the coordinate of the word w_i in dimension j of emoticon space. Thus, words are projected into emoticon space. By summing up the coordinates of the words that contained in a particular post, the projection from post to emoticon space is finished.

Then, the process of text feature extraction is done. Some supervised sentiment classification tasks can be performed by using the coordinates of the posts as text features.

4 Image Feature Extraction

As discussed in the paper, we not only consider the texts that contain emoticons, but also focus on the images uploaded in microblog posts. Usually, images are served as auxiliaries for texts. Therefore, based on the user action analysis, images should be regarded as important signals for expressing and conveying emotions. It is worthy to study how images convey emotions along with texts and play an important role in sentiment analysis in microblog environment.

In this section, there are two steps. First, we will explain the process of filtering images and image clipping. Second, we will introduce the method for extracting image features.

4.1 Filtering and Clipping

Usually, images have intensive relation to texts and thus reinforce the emotion expressed in texts. However, not every image is suitable for studying sentiment association with text. Advertisements and repeated posts are seen as spam microblogs for conveying invalid information. They should be filtered out. Considering images have different sizes, we clip the images to a specific size to prepare for later feature extraction.

4.2 Extracting Features

As mentioned in this paper, colors and textures are studied as sentiment features in existing works. Color histogram, a representation of the distribution of colors in an image, might indicates the emotion in it. For instance, an image with warm tone might represent positive sentiment. An image with cold tone might represent negative sentiment. The color histogram are often used for three-dimensional spaces like RGB and HSV.

To avoid complexity brought by high dimension, more advanced characteristics of image will not be introduced in our work. In microblog environment, we use naive method to acquire the color histograms of image built for RGB and HSV color space, which is proved relevant to emotion expressing. For instance, Fig. 2 shows an image with warm tone.

Fig. 2. An image with warm tone in microblog

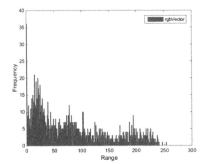

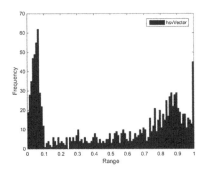

Fig. 3. RGB histograms of a warm-tone image

Fig. 4. HSV histograms of a warm-tone image

After clipping this image to 20 * 20 size, we acquire the histogram of RGB and HSV shown in Figs. 3 and 4.

As comparison, Fig. 5 shows an image with cold tone. The histogram of RGB and HSV are shown in Figs. 6 and 7.

Fig. 5. An image with cold tone in microblog

From Figs. 2, 3, 4, 5, 6 and 7 we can see the difference between images with warm and cold tones. The RGB and HSV histograms of image with cold tone are relatively mean.

Apart from histograms, we also make statistics about the amount of 16 specific colors contained in the pixels of image. The frequency of each of the 16 colors is seen as feature. Therefore, we acquire basic features of image. As introduced above, there are 371 values served as features, including 255 frequency values from RGB space, 100 frequency values from HSV space and 16 frequency values of specific colors appearance.

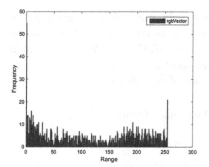

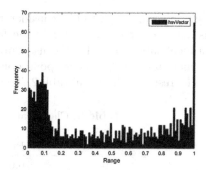

Fig. 6. RGB histograms of a cold-tone image

Fig. 7. HSV histograms of a cold-tone image

5 Experiments

5.1 Experiment Setups

Our experiments are performed on a Chinese microblog dataset that is crawled from Sina Weibo search engine using key words in food area. Each item in this dataset is a microblog post which contains both a text description and an image. After filtering out spam microblogs such as repeated posts and advertisements, we select about 3000 candidate microblogs for later training. The sentiment labels for microblogs are manually done by a crowdsource website. Each post is classified based on the emotional tendency. Positive emotions are defined as class 1, negative emotions are defined as class -1 while other neutral emotions are regarded as class 0. In annotation process, each post is labeled by two different persons. It is only when a posts label is agreed by two persons that the post is regarded to own a valid label. The statistic of training set is shown in Table 1.

Table 1. Proportion of each class

	Positive	Neutral	Negative	Total
Proportion	69 %	23 %	8 %	100 %

Since the amount of negative posts is relatively less compared with those of neutral and positive posts, we leave out the negative posts and only do classification task with positive posts and neutral posts (i.e. subjectivity classification).

5.2 Pre-experiment on Images

Before the formal experiment, we design a simple pre-experiment on images, which will help to observe the image features' influence from classification result. We pick a part of candidate microblogs to form a sub-dataset whose size is 324.

In pre-experiment, we only consider the images in microblogs. Each item of this sub-dataset contains 371 image features that is introduced above, and a sentiment label (0 or 1). We use Support Vector Machine (SVM) to train a classifier and test it on correspond double-fold validation set.

The result is shown in Table 2. The accuracy of classifier is 0.642.

Table 2. Classification based on image features

Class	Precision	Recall	F1-Score
0	0.69	0.58	0.63
1	0.60	0.71	0.65
Average/Total	0.65	0.64	0.64

We can see this classification performance is acceptable, which denotes images' features make sense in microblog sentiment classification. In normal experiment, we will perform subjectivity classification utilizing both textual and image features.

5.3 Subjectivity Sentiment Classification

For subjectivity classification, we first extract respective features of text and image, and then we use Support Vector Machine (SVM) to train classifiers. To extract text features, we select 100 most commonly used emoticons to construct emoticons space and make projection to it. Then we acquire 100 features for each text. For image features extraction, we get features directly from RGB and HSV histograms. Also, frequency of 16 specific colors are included as features. Thus, we acquire 371 features for each image.

Since negative posts are not considered, the original classification has actually become a binary-classification task or subjectivity task. Because of unbalanced proportion between neutral and positive posts, it may have influence on classifiers performance and thus cause low accuracy. To avoid this situation, we have balanced the training set and make the amounts of positive and neutral posts nearly the same. Considering the reliability of sentiment label annotated by crowdsource website, we only pick the most reliable part of candidate microblogs for training. After balancing, the amounts of positive and neutral posts are both 162.

Our experiment has two parts. For the first part, we only consider 100 textual features in microblog. Each microblog post in training set has 100 features and a sentiment label. We apply SVM to train a classifier called simple classifier and test it on correspond double-fold validation set. For the second part, besides the 100 textual features, we also incorporate 371 features of image in a microblog post. In this situation, each microblog in training set has 471 features and a sentiment label. We use SVM to train a classifier called complex classifier and test it on correspond double-fold validation set. For evaluation, we also randomly create a testing set, whose scale is 334, to verify the classification performance of complex classifiers with image features and textual features together.

5.4 Result Analysis

The experiment results are shown here. Table 3 indicates the first part, where we perform classification only considering the text features in training set.

Table 3. Classification based on texual features only

Class	Precision	Recall	F1-Score
0	0.55	0.55	0.55
1	0.50	0.49	0.50
Average/Total	0.52	0.52	0.52

Table 4 indicates the second part, where classification is performed based on synthesis of text and image features. Table 5 indicates the evaluation on testing set.

Table 4. Classification based on text and image features together

Class	Precision	Recall	F1-Score
0	0.59	0.56	0.57
1	0.54	0.56	0.55
Average/Total	0.56	0.56	0.56

Table 5. Classification on testing set

Class	Precision	Recall	F1-Score
0	0.61	0.54	0.57
1	0.58	0.64	0.61
Average/Total	0.59	0.59	0.59

In Table 3, the accuracy of classifier without image features is 0.525. While in Table 4, the accuracy of classifier with image features has increased to 0.562. Through accuracy comparison, the performance of complex classifiers is better. We make the hypothesis that the discrimination of two parts is not obvious. After significance test, the probability value of hypothesis is about $3.71*10^{-3}$ (far lower than 0.05), which proves improvement over baseline work shown in Table 3. The best classifier is linear SVM with parameter $C = 10$. The results from training set support our propose that in microblog environment, with the synthesis of text and image, classifiers have better performance in sentiment subjectivity classification task, than merely consider the text features. In testing set, the accuracy of complex classifier achieves at 0.591, which verifies and reinforces our conclusion.

In Fig. 8, Receiver Operating Characteristic (ROC) is plotted to visualize the performance of complex classifier on testing set. The X axis is false positive rate (FPR), while the Y axis is true positive rate (TPR).

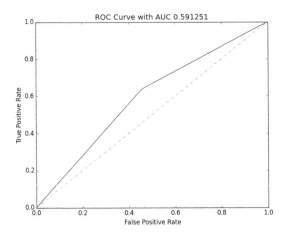

Fig. 8. ROC curve

From the figure, we can see this ROC curve is above the $y = x$, which indicates the benign performance.

6   Conclusion

In this paper, we discuss the sentiment analysis in microblog environment. Similar to previous research, we introduce machine learning method in sentiment classification task, but we also focus on the images in microblog posts. Compared with baseline works that mainly study textual features, our research incorporate image features in sentiment analysis. We extract image features from color histograms of RGB and HSV space. Also, frequency of specific colors are included as features. In experiment, our dataset is collected from Sina Weibo on area of food. We train two SVM classifiers for sentiment subjectivity classification task. Simple classifier is trained based on textual features, and the complex classifier is trained on image and text features together. Experiment results show that the performance of classifier with image features is better than simple classifier without image features on validation and testing set, which supports our proposal about synthetical consideration of text and image features in microblog sentiment analysis.

However, our research still need to be improved. In the future, we will try more advanced methods to extract image features besides histograms and frequency. Also, after image and textual features are mixed, we could perform PCA to acquire the most important and associated features to be used in training process. Besides, the performance of classifier might be influenced by the scale of dataset. Thus, we can adjust the dataset to find out how scale will influence the precision, recall and f1-score of classifier. Our dataset is built in food area. To prove our methods generality, we are trying to verify the benign performance on other areas such as economic, education and so on.

Acknowledgement. This work was supported by National Key Basic Research Program (2015CB358700) and Natural Science Foundation (61472206, 61073071) of China. Part of the work has been done at the Tsinghua-NUS NExT Search Centre, which is supported by the Singapore National Research Foundation & Interactive Digital Media R&D Program Office, MDA under research grant (WBS:R-252-300-001-490).

References

1. Go, A., Huang, L., Bhayani, R.: Twitter sentiment analysis. Entropy, 17 (2009)
2. Jiang, F., Liu, Y., Luan, H., Zhang, M., Ma, S.: Microblog sentiment analysis with emoticon space model. In: Huang, H., Liu, T., Zhang, H.-P., Tang, J. (eds.) SMP 2014. CCIS, vol. 489, pp. 76–87. Springer, Heidelberg (2014)
3. Kolchyna, O., Souza, T.T., Treleaven, P., Aste, T.: Twitter sentiment analysis. arXiv preprint arXiv:1507.00955 (2015)
4. Kouloumpis, E., Wilson, T., Moore, J.: Twitter sentiment analysis: the good the bad and the OMG!. ICWSM **11**, 538–541 (2011)
5. Liang, J., Chai, Y., Yuan, H., Zan, H., Liu, M.: Deep learning for Chinese microblog sentiment analysis. J. Chin. Inf. Process. **28**(5), 155–161 (2014)
6. Morency, L.P., Mihalcea, R., Doshi, P.: Towards multimodal sentiment analysis: harvesting opinions from the web. In: Proceedings of the 13th International Conference on Multimodal Interfaces, pp. 169–176. ACM (2011)
7. Pak, A., Paroubek, P.: Twitter as a corpus for sentiment analysis and opinion mining. In: LREC, vol. 10, pp. 1320–1326 (2010)
8. Pérez-Rosas, V., Mihalcea, R., Morency, L.P.: Utterance-level multimodal sentiment analysis. In: ACL (1), pp. 973–982 (2013)
9. Wu, Q., Zhou, C.-L., Wang, C.: Content-based affective image classification and retrieval using support vector machines. In: Tao, J., Tan, T., Picard, R.W. (eds.) ACII 2005. LNCS, vol. 3784, pp. 239–247. Springer, Heidelberg (2005)
10. Zhang, S., Yu, L., Hu, C.: Sentiment analysis of chinese microblog based on emoticons and emotion words. Comput. Sci. **39**(S3), 146–148 (2012)
11. Zhang, Y., Shang, L., Jia, X.: Sentiment analysis on microblogging by integrating text and image features. In: Cao, T., Lim, E.-P., Zhou, Z.-H., Ho, T.-B., Cheung, D., Motoda, H. (eds.) PAKDD 2015. LNCS, vol. 9078, pp. 52–63. Springer, Heidelberg (2015)

Nugget-Based First Story Detection in Twitter Stream

Yongqin Qiu, Sixu Li, Rui Li$^{(\boxtimes)}$, Lihong Wang, and Bin Wang

Institute of Information Engineering,
Chinese Academy of Sciences, Beijing 100190, China
{qiuyongqin,lisixu,lirui,wanglihong,wangbin}@iie.ac.cn

Abstract. Twitter First Story Detection (FSD) task refers to the detection of first tweet about the new event in tweet stream, which is a hard but important task in twitter event detection. However, the number of comparisons is too large in traditional online detection methods and there is a lack of global information of each event during detection process. To deal with the shortcomings above, we propose a novel FSD method based on nugget, which can describe the event concisely. Our approach generates and updates dynamically a nugget for each detected event in the process of detection. When a new tweet arrives, it is first compared with the nugget of each event, to be clustered into the event when it hits the nugget. Otherwise it is compared with individual tweets in the event. Our method improves the detection accuracy and reduces the number of comparisons. The experimental results on two public data sets show that our system has reached the state-of-the-art. Besides, we prove theoretically that our method possesses advantages in efficiency.

1 Introduction

With the rapid development of social networks, Twitter has become one of the most popular social network web sites with over 200 million active users, and 400 million tweets posted each day [1]. Some major events in Twitter could draw the public attention immediately and become a hot-spot. For this reason, Twitter has been used as a source of new event detection. Both public and private need not only to master the occurrence of new events in time, and also want to acquire the origin of the new events, namely, the first story.

First Story Detection (FSD) task in tweet stream is to identify the first tweet talking about a particular event and it is considered the most difficult task in TDT [5]. However a good FSD system would be crucial for government agencies and large enterprises to detect and analyze new event. The existing work basically used online clustering algorithm, which is called Single Pass, to handle this task. When a new tweet arrives, Single Pass algorithm computes the similarity between this tweet and all historical tweets, and clusters the new tweet to the bucket[1] with the most similar tweets. Much work has been done to improve the effective and efficiency, such as Yang [7], Zhang [6] and Petrovic [8].

---

[1] A bucket contains some tweets about the same event which is detected already.

© Springer Science+Business Media Singapore 2015
X. Zhang et al. (Eds.): SMP 2015, CCIS 568, pp. 74–82, 2015.
DOI: 10.1007/978-981-10-0080-5_7

However, to find the most similar tweet, traditional approaches needed to compute the similarity between new arrived tweet with all historical tweets, which did not considered the global information of detected events and needed a large number of comparisons. To overcome these shortcomings, we propose a Nugget-based first story detection method, which generates a quality nugget, a simple and effective description constituted by event information and metadata, for each bucket during Single Pass detection process. When a new tweet arrives, it is first compared with the nugget of each bucket. If it hits the nugget, it will be clustered to this bucket. If not, it is compared as traditional methods do. Experimental results on two public datasets (one is the biggest as we known) show that Nugget-based FSD method outperforms the state-of-the-art system currently and significantly reduces the false alarm and the number of comparisons during clustering process.

Our main contributions are: (i) we present a Nugget-based First Story Detection method in Twitter stream, which can improve the performance of detection and reduce the number of comparison than traditional methods. (ii) We practiced in the two public data sets in our experiments, one of which contains 81,087 Tweets and 505 events. To our knowledge it is currently the largest public data sets and it is also the first time used in Twitter FSD task.

2   Related Work

The problem of Topic Detection and Tracking (TDT) that ran as part of the Text REtrieval Conference (TREC) has been widely studied in traditional media [9]. In recent years, however, topic and event detection in Twitter has attracted much attention. Much research work focus on how to find emerging events, breaking news, and general topics that attract the attention of a large number of Twitter users.

Sankaranarayanan et al. [2] proposed a system, called TwitterStand, which employed a online clustering algorithm based on weighted term vector according to tf-idf and cosine similarity to form clusters of news. Becker et al. [10] used the classical incremental clustering algorithm, which has been proposed for NED in news documents, to identify of real-world event content and its associated Twitter messages, which continuously clusters similar tweets and then classifies the clusters content into real-world events or nonevents. In order to improve efficiency, Petrovic et al. [8] adapted variant of the locality sensitive hashing methods (LSH) [12], which limits the search to a small number of documents, to detect new events that have never appeared in previous tweets. Then Petrovic [13] used three sources of paraphrases and combine them with LSH to improve the effect of first story detection in Twitter. The results showed that their method can be very small C_{min} value (0.679), which is state-of-the-art as we know, and it is also the baseline we take.

The previous work which is similar to our work is summarization technology. Nichols [14] proposed an algorithm that generates a journalistic summary of an event using only status updates from Twitter as a source. Chakrabarti [15] formalized the problem of summarizing event-tweets and gave a solution based on learning the underlying hidden state representation of the event via Hidden Markov Models. Their model significantly outperforms some intuitive and competitive baselines through extensive experiments on real-world data. Popescu [16] adapted NLP techniques to automatically detect events involving known entities from Twitter and understanding both the events as well as the audience reaction to them. Yang [17] proposed a dynamic pattern driven approach to summarize data produced by Twitter feeds. The Nugget-based FSD method is somewhat analogous to summarization technology, which will generate high quality description namely "nugget" for each bucket during clustering process to assist FSD task.

3   Nugget-Based First Story Detection

3.1   Task Definition

Twitter First Story Detection (FSD) systems use new twitter stream as input, in which tweets are strictly time-ordered. Only previously received tweets are available when dealing with current tweet. The output of FSD systems is a decision for whether the current tweet is talking about a new event or not.

3.2   Nugget-Based FSD

The Nugget-based FSD algorithm we proposed is an improvement on the traditional Single Pass algorithm which can automatically generates a high-quality and representative nugget for each bucket during clustering process. When a new tweet arrives, it only needs to be compared to the nugget of each bucket, if it can hit the nugget of some bucket, it will be clustered to the bucket and updates the nugget. Otherwise, it will be compared to some of the tweets in each bucket to decide which bucket to put in. If the similarity between the new tweet and every bucket is sufficiently low, it will be judged as a first story of a new event. The frame-work of summarization-based FSD algorithm is shown in Algorithm 1.

3.3   Nugget Generation and Update

As described above, the key point of nugget-based FSD algorithm is the generation and update of nugget, which determines the effect of the algorithm. Nugget is composed by two parts: (i) Entity information extracted from knowledge base; (ii) Metadata: hashtag and at ("@") information, the hashtag included in tweets is a good indicator to event and at ("@") information will include related user about some event. The following will describe in detail how they are generated.

Entity Information: Intuitively, extracting entity information of a tweet from knowledge base can be very useful to describe what the tweets mainly talk about. We adopt the approach proposed by Meji [18] to detect the entity of a twitter from Wikipedia. The method uses an entity linking relationship to

---

Algorithm 1. summarization-based FSD Algorithm

Input:
    tweet stream $\mathbb{C}$, threshold θ
Output:
    every tweet labeled "first story" or "old story"
1: **for** each $t_i \in \mathbb{C}$ **do**
2:     generate summarization t_{i_sum} for t_i
3:     **if** the set of buckets $\mathbb{B} = \phi$ **then**
4:         create a new bucket B_1
5:         $B_1 = B_1 \bigcup t_i$
6:         generate summarization B_{1_sum} for B_1
7:         label "first story" to t_i
8:     **else**
9:         compare t_{i_sum} with each B_{j_sum} (B_{j_sum} is summarization of B_j, $B_j \in \mathbb{B}$)
10:        **if** t_{i_sum} hit B_{j_sum} **then**
11:           $B_j = B_j \bigcup t_i$
12:           update B_{j_sum}
13:           label "old story" to t_i
14:        **else**
15:           Let maxS=$max_j(\text{sim}(t_i, B_j))$, j=1,2,...,K
16:           **if** $maxS \geq \theta$ **then**
17:             $B_j = B_j \bigcup t_i$
18:             update B_{j_sum}
19:             label "old story" to t_i
20:           **else**
21:             create a new bucket B_{new} in $\mathbb{B}$
22:             $B_{new} = B_{new} \bigcup t_i$
23:             generate summarization B_{new_sum} for B_{new}
24:             label "first story" to t_i
25:           **end if**
26:        **end if**
27:     **end if**
28: **end for**

---

gather Wikipedia entries that are semantically related to a tweet: the common-ness probability is based on the intraWikipedia hyperlinks, which computes the probability of a concept/entity c being the target of a link with anchor text q in Wikipedia by:

$$commonness(c, q) = \frac{|L_{q,c}|}{\sum_{c'} |L_{q,c'}|} \tag{1}$$

where $L_{q,c}$ denotes the set of all links with anchor text q and target c.

    Tweets are then represented as the bag-of-entities derived from linking each n-gram in the content of the tweet to the most probable Wikipedia entity. In case of n-gram overlap, only the longest is considered.

Metadata: Metadata is an important indicator to event detection. Event-oriented target tweets can be identified by a few words or metadata as mentioned by Spina [19]. This discovery inspires us to use metadata such as hashtag and "@" as one part of our nugget to handle FSD task.For example, in our corpus, the event "NBA between Bucks and Celtics" includes 2 tweets: one is "Ive never seen the Celtics play this terribly #enoughsaid", and anther is "Bucks are winning #enoughsaid and that even my teammm", we can easily put them to the same bucket rather than compare it with every tweets in the bucket, they are likely to discuss the same topic because have the same hashtag #enoughsaid. At information includes the related user about some event, it is also helpful to detect the tweets talk about the same event.

Nugget Update: With the incremental of the tweets in every bucket, the number of nugget is increasing. If we don't update the nuggets in real time and choose the quality nugget in the bucket, there could be more noise and harmful to the results. Therefore we introduce KL distance to pick the most representative nugget. KL divergence is defined as follows [20]:

$$KL(P||Q) = \sum_x p(x) log \frac{p(x)}{q(x)} \tag{2}$$

The more a term occurs within the event, and the less it occurs in other event, the higher the weights it should be assigned.

4   Experiments

4.1   Dataset and Evaluation

Datasets: We carry out our experiment on two datasets. One is the same as [13], which we call *data1*, including 2363 tweets and 27 events labeled. There are 3034 tweets and 27 events in the original dataset, but about 22 % was lost when we crawl it from twitter API. In order to avoid occasionality, we use another dataset [1], *data2*, including 81087 tweets and 505 events, which is, to our knowledge, the largest public dataset available for FSD task in Twitter.

Evaluation: Unless stated otherwise, we use the official evaluation metrics and parameters from TDT^2: *miss probability*, P_{miss} and *false alarm*, P_{FA}. C_{det} is computed as [8].

$$C_{det} = C_{miss} * P_{miss} * P_{target} + C_{FA} * P_{FA} * P_{non-target} \tag{3}$$

where C_{miss} and C_{FA} are costs of miss and false alarm, P_{target} and $P_{non-target}$ are the prior target and non-target probabilities. C_{min} is the minimal value of C_{det} over all threshold values.

---

^{2} ftp://jaguar.ncsl.nist.gov/current_docs/TDT3eval/TDT3fsd.pl.

4.2   Results and Analysis

We compare our approach with paraphrased-based method [13], to our knowledge which is the state-of-the-art both in effect and efficiency results for FSD task. This system uses LSH to reduce the time consuming and uses paraphrases to alleviate the high degree of lexical variation in documents that talk about the same event.

Experiment 1 Comparison with State-of-the-Art: Table 1 shows the results of C_{min} on $data1$. we use 'concept+hashtag+@' information to generate nuggets. We can see that our Nugget-based FSD method significantly reduce C_{min} by 39.6 %. Table 2 shows the detail performance of our nugget-based method on P_{miss}, P_{FA} and C_{det}. The C_{det} of our Nugget-based FSD method is lower than 0.679 over all threshold. Besides, the value of P_{FA} is always within a small range whatever the threshold is. Because old story can be easily confirmed via nugget and similarity comparing is avoided.

Table 1. C_{min} of two systems

System	Cmin
LSH-baseline	0.679
Nugget-FSD	**0.410**

Table 2. Detail of performance

threshold	P_{miss}	P_{FA}	C_{det}
0.10	0.518	0.005	0.566
0.15	0.481	0.004	0.514
0.20	0.444	0.006	0.495
0.25	0.333	0.009	**0.414**
0.30	0.333	0.012	0.441
0.35	0.333	0.013	0.450

Experiment 2 Effect of Different Signals: In this set of experiments, we test how the different signals in nugget effect the detection performance. Figure 1 shows the result of 4 different combination of 3 kind of signals. "entity" means using Entity information extracted from Wikipedia only; "entity+@" means using wikipedia entity and twitter at informaton; "entity + hashag" means using wikipedia entity and hashtag information. As can be seen from the figure:

First, the curve of 'entity+hashtag' is under 'entity', which means hashtag play a positive role.

Second, the P_{miss} result of 'entity' is only better than 'entity+@', and 'entity+ hashtag' is better than 'entity+hashtag+@', which means the information '@' in tweet leads to a negative effect. The result is different from our original expectation, many of them refer to official Twitter accounts like 'Reuters', 'ABC News', or 'BBC News', etc. If a first story mentions such a named user, it will become a miss case.

Lastly, all 4 methods have a lower bound of P_{miss} at 0.25 and upper bound of P_{FA} at 0.025, which is different from traditional FSD's result which both P_{miss} and P_{FA} fall in between 0 and 1.

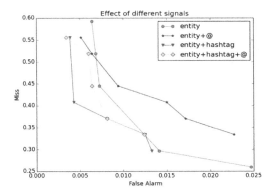

Fig. 1. Effect of different signals

Experiment 3 Validated in the Largest Public Data Sets: We run our method on *data2*, the largest public data with labeled information. We run it on 3000 tweets with 24 events, 10000 tweets with 64 events and 81087 tweets with 505 events respectively. Detailed performance is showed in Table 3 which shows our approach on *data2* achieve the comparable effect with on *data1*. We can also see from Table 3 that P_{miss} becomes higher and P_{FA} becomes lower with more testing data i.e. new first stories. Obviously, as the number of nugget increases, the probability of a tweet hitting a certain nugget also increases. Therefore, a new first story will be more likely to be missed and an old story will be more likely to be detected.

4.3   Time Analysis

In this section, we theoretically analyze the time complexity of LSH-FSD method and our Nugget-FSD method. It is worth mentioning that the LSH method [8] is the state-of-the-art in efficiency.

The time complexity predicting a story is new or old by LSH method can be computed as:

$$T_{hash} = P_{call} \cdot N \cdot t_{one} \tag{4}$$

P_{call} is the probability of two story x and y colliding using [8]. N is the number of historical data in database. $t_{one} = c$ is the time of computing two tweets' similarity.

The time complexity of our method consists of three parts. (i) the time of comparing nugget; (ii) comparing historical stories if it fails to hit the bucket's nugget; (iii) time of updating nugget.

$$T_{p\_s} = T_{nugget} + (1 - P_{nugget}) \cdot T_{hash} + T_{update} \tag{5}$$

$$T_{nugget} = O(t_{one}), T_{update} = len \cdot O(1) \tag{6}$$

$$P_{nugget} = P_{non-target} < 1, 1 - P_{nugget} = P_{target} = o(1) \tag{7}$$

We can use LSH method to hash every buckets nugget, and the time of T_{nugget} can be calculated as tone approximately. P_{nugget} means the probability

Table 3. Performance of Nugget-FSD in data2

threshold / nums of tweets	3 000(24)		10 000(64)		81 087(505)	
	P_{miss}	P_{FA}	P_{miss}	P_{FA}	P_{miss}	P_{FA}
0.10	0.625	0.009	0.640	0.006	0.680	0.004
0.14	0.425	0.010	0.563	0.007	0.602	0.006
0.18	0.30	0.013	0.469	0.009	0.521	0.007
0.22	0.25	0.021	0.422	0.013	0.473	0.009
0.26	0.075	0.025	0.312	0.018	0.388	0.012
0.30	0.100	0.037	0.219	0.028	0.295	0.017
0.34	0.075	0.053	0.187	0.041	0.210	0.025
0.38	0.075	0.058	0.172	0.045	0.202	0.031

of one tweets nugget hitting one of the buckets nugget. Ideally, all old tweet can hit a certain nugget. The third part is shown in Eq. 3.1. len is the length of one tweet, and len is usually short. So we have:

$$T_{P\_S} = T_{summarization} + P_{target} \cdot T_{hash} + T_{update}$$
$$= O(t_{one}) + P_{target} \cdot T_{hash} + l \cdot O(1)$$
$$= o(T_{hash}) + O(c) \tag{8}$$

From the above analysis we can see that our Nugget-based FSD method shows lower time complexity than LSH-based method. It can be more significant when used in a large amount of data or small P_{target} case.

5 Conclusion

In this paper we propose a novel Twitter first story detection method based on nugget. Our approach generates and updates nugget dynamically for each detected bucket in the process of event detection. When a new tweet arrives, it is compared with the nugget of each bucket, rather than compared with individual tweets in the bucket. The experimental results on two public data sets show that our system has reached the state-of-the-art. Besides, we prove theoretically that our method possesses advantages in efficiency. In the future, we will focus on how to generate high quality nugget and optimize the strategy of updating the nugget.

References

1. McMinn, A.J, Moshfeghi, Y., Jose, J.M.: Building a large-scale corpus for evaluating event detection on Twitter. In: Proceedings of the 22nd ACM International Conference on Information and Knowledge Management, pp. 409–418. ACM (2013)
2. Sankaranarayanan, J., Samet, H., Teitler, B.E. et al.: Twitterstand: news in tweets. In: Proceedings of the 17th ACM SIGSPATIAL International Conference on Advances in Geographic Information Systems, pp. 42–51. ACM (2009)

3. Allan, J.: Topic Detection and Tracking: Eventbased Information Organization. Kluwer Academic Publishers, Norwell (2002)
4. TDT 2004: Annotation manual (2004)
5. Allan, J., Lavrenko, V., Jin, H.: First story detection in TDT is hard. In: Proceedings of the Ninth International Conference on Information and Knowledge Management, pp. 374–381. ACM (2000)
6. Zhang, K., Zi, J., Wu, L.G.: New event detection based on indexing-tree and named entity. In: Proceedings of the 30th Annual International ACM SIGIR Conference on Research and Development in Information Retrieval, pp. 215–222. ACM (2007)
7. Yang, Y., Pierce, T., Carbonell, J.: A study of retrospective and on-line event detection. In: Proceedings of the 21st Annual International ACM SIGIR Conference on Research and Development in Information Retrieval, pp. 28–36. ACM (1998)
8. Petrovic, S., Osborne, M., Lavrenko, V.: Streaming first story detection with application to Twitter. In: Human Language Technologies: The 2010 Annual Conference of the North American Chapter of the Association for Computational Linguistics, pp. 181–189. Association for Computational Linguistics (2010)
9. Allan, J.: Introduction to topic detection and tracking. In: Allan, J. (ed.) Topic Detection and Tracking. The Information Retrieval Series, vol. 12, pp. 1–16. Springer, US (2002)
10. Becker, H., Naaman, M., Gravano, L.: Beyond trending topics: real-world event identification on Twitter[J]. In: ICWSM, vol. 11, pp. 438–441 (2011)
11. Allan, J., Papka, R., Lavrenko, V.: On-line new event detection and tracking. In: Proceedings of the 21st Annual International ACM SIGIR Conference on Research and Development in Information Retrieval, pp. 37–45. ACM (1998)
12. Gionis, A., Indyk, P., Motwani, R.: Similarity search in high dimensions via hashing. In: VLDB, vol. 99, pp. 518–529 (1999)
13. Petrovic, S., Osborne, M., Lavrenko, V.: Using paraphrases for improving first story detection in news and Twitter. In: Proceedings of the 2012 Conference of the North American Chapter of the Association for Computational Linguistics: Human Language Technologies, pp. 338–346. Association for Computational Linguistics (2012)
14. Nichols, J., Mahmud, J., Drews, C.: Summarizing sporting events using Twitter. In: Proceedings of the 2012 ACM International Conference on Intelligent User Interfaces, pp. 189–198. ACM (2012)
15. Chakrabarti, D., Punera, K.: Event summarization using Tweets[J]. In: ICWSM, vol. 11, pp. 66–73 (2011)
16. Popescu, A.M., Pennacchiotti, M., Paranjpe, D.: Extracting events and event descriptions from Twitter. In: Proceedings of the 20th International Conference Companion on World Wide Web, pp. 105–106. ACM (2011)
17. Yang, X., Ghoting, A., Ruan, Y. et al.: A framework for summarizing and analyzing Twitter feeds. In: Proceedings of the 18th ACM SIGKDD International Conference on Knowledge Discovery and Data Mining, pp. 370–378. ACM (2012)
18. Meij, E., Weerkamp, W., de Rijke, M.: Adding semantics to microblog posts. In: Proceedings of the Fifth ACM International Conference on Web Search and Data Mining, pp. 563–572. ACM (2012). McMinn, A.J., Moshfeghi, Y., Jose, J.M
19. Spina, D., Gonzalo, J., Amig, E.: Learning similarity functions for topic detection in online reputation monitoring. In: Proceedings of the 37th International ACM SIGIR Conference on Research Development in Information Retrieval, pp. 527–536. ACM (2014)
20. Kullback, S., Leibler, R.A.: On information and sufficiency. Ann. Math. Stat. **22**, 79–86 (1951)

Dirichlet Process Mixture Model
for Summarizing the Social Web

Xinjun Guan, Ying Yang, Xinru Yang, and Chen Lin[✉]

Department of Computer Science, Xiamen University, Xiamen, China
chenlin@xmu.edu.cn

Abstract. Automatic summarizations have gained increasing attentions
as they not only improve reading experiences but also facilitate management of collective knowledge on the social web. The social web is featured
by social interactions. Ignoring this type of information limits the ability of traditional summarization techniques to generate more intelligent
and comprehensive summaries. In this paper we present a mixture model
based on Dirichlet Process, which exploits information contained in tags
and other social behaviors. The model assigns each sentence one explicit
"topic". The assignment follows a Chinese Restaurant Process, where an
infinite number of topics are organized by a tag or group. The model
has straight-forward applications to diverse social summarization tasks.
It is a natural fit for flexible data structures and incremental compu-
tations. We present applications to tag-driven summarization, comparative summarization and update summarization. We evaluate our model
through both quantitative and qualitative experiments on various real
world data sets.

Keywords: Automatic summarizations · Social web · Dirichlet process
mixture model

1 Introduction

We are in the age of social web, an ecosystem where everyone contributes to the
collective knowledge. The collective knowledge comes from both user-published
contents (i.e. posts, tweets, and blogs), and user interactions (i.e. tagging, reply-
ing and commenting). Such contents and interactions can be observed in communities with various purposes, including news, QA and so on. Within each
community, the textual contents and interactions are appealing data sources to
mine the insights. However, mining the social web data is difficult, due to its
extremely high volume. Automatic summarization is usually necessary.

Text summarization is not a new problem. There are fruitful summarization
systems that generate summaries by exacting and combining several key sentences.
The key sentences are often determined by a global ranking mechanism that computes the capacity of each sentence to embrace similar sentences [4, 7, 15]. Although
global ranking by textual similarities has been validated extensively on pure texts,
in the domain of social web, its ability is limited as it completely ignores the "social
aspect" of social web.

© Springer Science+Business Media Singapore 2015
X. Zhang et al. (Eds.): SMP 2015, CCIS 568, pp. 83–94, 2015.
DOI: 10.1007/978-981-10-0080-5_8

The social aspect plays an important role in utilizing the social web. On one hand, social interactions provide valuable information. User generated contents are in general brief and informal, which makes the inference of "topics" inaccurate and unreliable. Replies, comments and the main body of posts complement each other to form a group of related contents. Tags operate in a more obvious way to group contents created by different users. This grouping information is helpful in inferring the "topics". On the other hand, the social nature poses challenges, in the sense that people need more intelligent and comprehensive summarizations. In the past, traditional summarization systems work on a single document or a collection of documents. Nowadays, summarization are demanded in diverse scenarios. We next show a few possible examples.

Tag-Driven Summarization. People use two types of tags to annotate contents, tags that are recommended by the system, or terms that they have coined by themselves. In both cases, the meanings of tags are unclear and broad, especially for invented tags that never exist before. Therefore, it is beneficial to offer a tag-driven summarization, which covers possible "topics" that explain the tag, based on tagged contents. It is of utmost importance when a new term emerges and becomes popular. Tag-driven summarization is not a trivial task. In most cases, multiple tags are associated with one piece of contents. In order to produce a precise summary, the delicate relationships that which part of contents is related to which topic under which tag, need to be uncovered. A similar application is **group summarization**, when a group of people is involved. A group can be regarded as a tag, and corresponds to a set of standpoints.

Comparative Summarization. A further implementation is comparative summarization for two entities (i.e. document, tag, group of people). Comparative summarization draws the most informative skeleton out of redundant contents, including the common and distinguishing parts of the two entities. Comparative summarization is only possible when entity-specific topics are identified.

Update Summarization. A social web site is in a consistently changing state. There are new users joined, new contents created, and new topic and groups formed in every minute. The dynamic nature of social web requires an online fashion of the update summarization system. The essentials of update summarization are two-fold. (1) A solution that incrementally builds a model according to the new data, without the whole corpus. (2) A framework that catches new "topics" on the fly, while keeping track of the old "topics".

Previous summarization methods do not apply well to social web data, as they lack explicit modeling of "topics" with respect to social interactions. One alternative is to turn to topic modeling approaches with supervision of tags. Most researches in literature adopt a LDA [2] type of probabilistic models to generate a fixed number of topics [1,6,8]. This assumption is not reasonable for a social web site where there could be a countless number of "topics". For example, in an online debate forum, participants can either support an existing point of view, or express novel opinions. Restricting the model to a predefined level of abstraction (i.e. fixed number of topics) seems to be improper. Furthermore, they are incapable of generating update summaries in an incremental manner.

There is a recent trend in adopting Dirichlet Process in topic modeling [3,14]. The clustering property of the Dirichlet process serves as a nonparametric prior for the number of topics. We follow this path with the following improvements.

(1) We assume that for a text segment concise enough, only one topic is expressed. Social texts are usually short and sparse. By assigning a topic label to a text segment instead of a word, we make full use of the information within a reasonable window of the target word, thus outcome the problem of sparsity and high dimensions. This assumption also makes it more natural to select sentences from a topic.
(2) We consider a hierarchical model, where the posting/replying/commenting behavior of a tag/group is a Chinese Restaurant Process (CRP). The adoption of CRP is advantageous in three ways. (1) CRP can automatically estimate the number of topics. It is more flexible as it "lets the data speak". (2) CRP demonstrates a phenomenon of "rich get richer", which is a vivid simulation of how topic evolves in real social networks. (3) CRP enables a straight-forward inference of new topics, which can be conducted in update summarization.
(3) We incorporate a background topic to purify tags/groups. The background topic is worthy of inclusion because it filters non-informative words. It can also be viewed as a common topic in comparative summarization.

We show three direct applications of this model, including tag-driven summarization, comparative summarization and update summarization. We verify the model's competency in various real world data sets. We find that our model can achieve significantly better performances, compared with state-of-the-art approaches.

The rest of the paper is organized as follows. Section 2 introduces related works on multi-document summarization and topic models. Section 3 overviews the model framework and derives a Gibbs Sampling algorithm to learn topic assignments. Section 4 analyzes experimental results and examples on different real data sets. Section 5 presents our conclusion.

2   Related Work

Multi-document summarization is a well-studied area. The goal of multi-document summarization is to convey the main and most important meaning of several documents. Previous works on multi-document summarization fall into two categories, extractive methods and abstractive methods. The former class selects and combines representative sentences to form the summary. Most common selection criteria include Lexrank [4], structural centroid in a sentence graph; or DSDR [7], degree of approximation to reconstruct other sentences; or other types of information quality in a cluster [13]. The latter class, which fills phrases in a predefined template is less adopted.

Although extractive methods are so far the most successful paradigm in multi-document summarization, existing extractive methods are not perfect for social web data. The main reason is that they do not employ social interactions. Missing the information contained in social interactions, existing methods apply global ranking to the corpus as a whole, leading to a vague summary whose coverage over the tag space (and other social entities) can not be measured. Moreover, they do not have an explicit definition of "topic", thus are more sensitive to observed data, and can not adjust to future data.

Topic models are probabilistic graphical models for text clustering. LDA [2] is an exemplification of topic models. In LDA and its many variants, the number of clusters is fixed a priori. This may raise questions when credible prior knowledge is not available. To address these problems, nonparametric models are presented, most of which are based on Dirichlet Process (DP). For example, HDP [14] and HLDA [3] DP based model that learn hierarchical topic structures. Instead of allowing the parameters to "grow" as more data is observed, some researchers walk in an opposite direction. GSDMM [16] trims the number of clusters by first setting it to a maximal number of data points, then letting it shrink to converge. Apparently, a significant disadvantage of GSDMM is its incapability to increase the size of the model when new data arrives.

Though topic models are mainly used in text clustering, supervision can be incorporated, such as LabeledLDA [11]. LDA-type models have achieved satisfying performances on summarization tasks [1,6]. However, most of them assign a topic variable to each word, which makes the selection of sentences complicated and indirect.

3   Model

3.1   Chinese Restaurant Process Model

We first introduce the preliminaries. In most social networking sites, we observe similar structures. When a user publishes a piece of contents, other users might comment on it, or reply to it. The social interactions triggered by a single piece of contents lead to a collection of contents, including the original posts, replies, comments, responses to comments and so on. This collection of contents is denoted as a "document" d. A "document" is usually labeled by multiple tags, which we refer to as $b \in B$. For instance, tweets are labelled by hashtags, questions in Quora are labelled by topics. Note that tags are shared among all units in a "document". The unit text segment in the "document" is regarded as u, which can be a sentence in the original post, a title, or a short comment. A topic with indicator a is a probability distribution β_a over words w. The topics are organized by tags. Each tag is associated with a set $a(b)$ of infinite topics.

We assume that in each unit text segment, one and only one topic is expressed. Let's consider the simplified case where each user joining in a discussion only writes one sentence. (The case where users write long posts is by no means different.) Intuitively, in each sentence, the user chooses to elaborate on one topic that corresponds to one tag. As an ordinary person, the user is gregarious, which

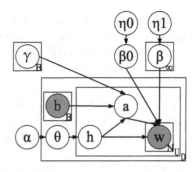

Fig. 1. Graphic representation of the Chinese Restaurant Process Model

means that he is more likely to choose a popular topic among all existing ones. He is less likely, not entirely impossibly to stand out and declare a new standpoint. Once the topic is determined, the user chooses words from the vocabulary related to his topic. From time to time, the user needs some non-informative sentences as conjunctives. These words are usually related to what is talked about in general in the outer space (i.e. in the whole forum), but not specifically related to the given topic. These intuitions are modeled in our CRP model.

The generative process for CRP, as illustrated in Fig. 1, is listed below.

- For each topic $a = 1 \sim \infty$
  - Draw a vocabulary $\beta_a \sim Dirichlet(\eta_1)$
- Draw a vocabulary for background topic $\beta_0 \sim Dirichlet(\eta_0)$

  For each document d

- Draw a switch distribution $\theta \sim Dir(\alpha)$. θ is a vector, whose dimension is the number of tags $\theta \in R^{|B|}$.
- For each unit text segment u
  - Draw a switch $h \sim Mul(\theta)$. h indicates which tag is chosen, $0 \le h \le |B|$.
  - Draw a topic $a|h, b, \gamma$. If $h = 0$, the topic is a background topic $a = 0$. Otherwise, the topic is drawn from a Chinese Restaurant Process from the topic sets $a(b_h)$ associated with the h-th tag b_h. The Chinese Restaurant Process $a_u \sim CRP(\gamma_h)$ states that, the probability of drawing from an old topic z is proportional to the population of topic $\frac{n_z}{n+\gamma_h}$, while the probability of generating a new topic z' is proportional to the scaling parameter $\frac{\gamma_h}{n+\gamma_h}$.
    * For each word $i = 1 \cdots N$
      · If $h_i = 0$, draw a word $w_i \sim Mult(\beta_0)$
      · Else if $h_i! = 0$, draw a word $w_i \sim Mult(\beta_{a_u})$.

3.2 Algorithm

In the Gibbs Sampling framework, We need to estimate the joint probability of the states of hidden variables, including the topic a_u and the switch h_u for each unit text segment u, given the states of all other variables. Parameters β, g, θ are marginalized. For computational convenience, we suppose the sets of topic indicators $\{a(b)\}$ for all tags b are disjoint, $a(i) \cap a(j) = \emptyset, \forall i \neq j$. Suppose in the previous iteration, the maximal indicator of topic is K, we have:

$$p(h_u = s, a_u = z | \boldsymbol{h}_{-u}, \boldsymbol{w}, \boldsymbol{a}_{-u}, b) \tag{1}$$

$$\propto \begin{cases} p(h_u = s | \boldsymbol{h}_{-u}) p(a_u = z | h_u = s, \boldsymbol{a}_{-u}, b) p(\boldsymbol{w}_u | a_u = z, \boldsymbol{w}_{-u}), & s \neq 0, 0 < z \leq K+1 \\ p(h_u = 0 | \boldsymbol{h}_{-u}) p(\boldsymbol{w}_u | h_u = 0, \boldsymbol{w}_{-u}), & s = 0, z = 0 \end{cases}$$

We omit the hyper-parameters α, γ, η for the limited space. The first term is the posterior probability of the switch is $p(h_u = s | \boldsymbol{h}_{-u}, \alpha) = \frac{\alpha_s + \sharp[h_{d,-u} = s]}{\sum_{s=0}^{|B|} \alpha_s + |U_{d,-u}|}$, where $|U_{d,-u}|$ is the number of segments in document d excluding the current one, $\sharp[h_{d,-u} = s]$ is the number of times a segment other than u links to the switch s. The second term is the Chinese Restaurant Process. For simplification, we let $f_z = p(\boldsymbol{w}_u | h_u = s, a_u = z, \boldsymbol{w}_{-u}, \eta), 0 < z \leq K$ to denote the probability of generating the current word vector given topic assignment z. f_z can be computed by directly expanding the Dirichlet probability [16] $f_z = \frac{\Pi_{w \in U} \Pi_{i=0}^{\sharp[w_u = w]-1} (\eta_z^w + \sharp[a_{-u} = z, w_{-u} = w] + i)}{\Pi_{j=0}^{|u|-1} (\Sigma_w \eta_z^w + \sharp[w_{-u} = z] + j)}$, in which $\sharp[a_{-u} = z, w_{-u} = w]$ is the number of times word w is assigned to topic z in segments other than the current one u, $\sharp[w_{-u} = z]$ is the number of times a word in other segments assigned to the current topic, $\sharp[w_u = w]$ is the number of times word w appears in the current segment, $|u|$ is the length of the current segment. We use the prior distribution to estimate the probability of generating the observed contents for a new topic.

$$f_{K+1} = \frac{\Pi_{w \in U} \Pi_{i=0}^{\sharp[w_u = w]-1} (\eta_z^w + i)}{\Pi_{j=0}^{|u|-1} (\Sigma_w \eta_z^w + j)}.$$

The Gibbs Sampling algorithm is summarized at Algorithm 1. In practice, we do not initialize $a_u, h_u, \forall u \in U$ randomly. On the contrary, we randomly sort the text segments, and initialize their states according to Eq. 1, where the observed words $\boldsymbol{w}_{-u}$ are the ones previously initialized. In each probability computation, we need the products of all words in the same segment. To avoid overflow, we compute the logarithm of probability for each word, and rolls the range of the natural exponential function of the aggregated log-probability in roulette wheel selection. After burn in, we compute the desired parameters β as:

$$\beta_a^i = \frac{\sharp[w = i, a = a] + \eta_1}{\Sigma_j \{\sharp[w = j, a = a] + \eta_1\}} \tag{2}$$

$$\beta_0^i = \frac{\sharp[w = i, a = 0] + \eta_0}{\Sigma_j \{\sharp[w = j, a = 0] + \eta_0\}}. \tag{3}$$

3.3   Application

We finally introduce three straight-forward applications of the CRP model, namely tag-driven summarization, comparative summarization and update sum-

---

Input: $w_u, \forall u \in U, b$
Output: $a_u, h_u, \forall u \in U$
1 **while** *not converged* **do**
2   **for** $u \in U$ **do**
3   Update all counters by excluding the current segment;
4   Delete z where $\sharp[a = z] = 0$;
5   Rearrange all topics to be a succession of topic indicators;
6         $K \leftarrow \max z$;
7   **for** $0 \leq z \leq K$ **do**
8           $f_z = \dfrac{\Pi_{w \in U} \Pi_{i=0}^{\sharp[w_u = w] - 1}(\eta_z^w + \sharp[a_{-u} = z, w_{-u} = w] + i)}{\Pi_{j=0}^{|u| - 1}(\Sigma_w \eta_z^w + \sharp[w_{-u} = z] + j)}$;
9   **end**
10        $f_{K+1} = \dfrac{\Pi_{w \in U} \Pi_{i=0}^{\sharp[w_u = w] - 1}(\eta_z^w + i)}{\Pi_{j=0}^{|u| - 1}(\Sigma_w \eta_z^w + j)}$;
11   **for** $0 < s < |B|, 0 < z \leq K + 1$ **do**
12          $p(h_u = s, a_u = z) = \dfrac{\sharp[a_{-u} = z]}{\sharp[b_{-u} = b_{d,s}] + \gamma} \dfrac{\alpha_s + \sharp[h_{d,-u} = s]}{\Sigma_{s=0}^{|B|} \alpha_s + |U_{d,-u}|} f_z$;
13   **end**
14   Sample $p(h_u, a_u)$;
15   Update all counters by including the current segment;
16   **end**
17 **end**

Algorithm 1. Gibbs Sampling for CRP

marization. For a given tag b, the tag-driven summarization is generated as follows. In Algorithm 1, we gather the set of the topics $a(b)$ related to b, and sentences that are assigned with $a \in a(b)$. For each tag-specific topic $a \in a(b)$, we select one representative sentence, which is the top sentence by a ranking formula, to build a candidate set $U(b)$. We will further analyze the performance of different rankings in the experiment section. Then we rank each sentence in the candidate set $u \in U(b)$ by the popularity of the topic, which is the number of sentences assigned with it. We combine the top sentences until the maximal length of summary is obtained.

The comparative summarization is implemented a little differently. Consider the problem of producing comparative summarization for two tags i and j, we first obtain documents that are only labelled by i as the tag-specific corpus $D(i)$, and similarly documents that are only labelled by j as the tag-specific corpus $D(j)$. We merge the two corpus as D. We then run Algorithm 1 on D, with one background topic, and two tags $|B| = 2$. It is easy to see that, the background topic is the common topic that both tags i, j have, the specific aspects of tag i are the topics linking to $a(i)$, the specific aspects of tag j are the topics linking to $a(j)$. As in tag-driven summarization, the assignments by Algorithm 1 are used to generate comparative summary. We choose and combine representative sentences from the background topics as the "commons" in the comparative summary, key sentences from the tag-specific topics as the "differences".

In the setting of update summary, we are facing the problem of updating the model for a modified corpus $D' = D_{new} + D_{old}$, given that we have already learnt the model for the old corpus D_{old}. The key point is to avoid re-computation for the whole corpus, while keeping the ability to detect emerging topics. CRP is built with a mechanism to learn new topics. We conduct Algorithm 1 directly on the new corpus D_{new}. Note that for old words, all the counters in the old corpus are added to counters in the new corpus. For example, the topic-word counter $\sharp[w_{-u} = w, a_{-u} = z]$ in the new corpus, is equal to $\sharp[w_{u'} = w, a_{u'} = z, u' \in D_{old}] + \sharp_{D_{new}}[w_{-u} = w, a_{-u} = z]$. If a word has never appeared in any of the old documents, we do not change its counters.

4   Experiment

In the experiments, we evaluate the performance of the proposed CRP model. We use four real world data sets. The DUC2007 data set (DUC) and the Reuters 21578 data set (Reuter) are collections of news reports. The Zhihu data set is a crawl from a large Chinese QA community (zhihu.com). Starting from a portal page (http://www.zhihu.com/topic/19760570) containing all discussions related to questions in the form of "how do people think about xxx", we crawl all the questions, answers, comments to answers and replies to comments. The Healthcare data set consists of short responses obtained in a telephone survey for Obama's healthcare plan [9]. All the four data sets are tagged. In the DUC and Reuter data sets, the tags are the topic labels assigned with each report. In Zhihu data set, the tags are associated with each question. We assume answers for the question share the tags. In the Healthcare data set, there are two tags, "support" and "against". More details of the data sets are illustrated in Table 1.

Table 1. Statistics of Data sets

Metric	DUC	Reuter	Zhihu	Healthcare
#documents	1240	1872	94783	942
#tags	50	117	1503	2
#tag per doc	1.01	2.57	3.90	1
#docs per tag	24.8	41.07	246.02	471

In pre-processing, we do not remove stop-words. Porter stemmer is adopted in indexing. But the stemmed terms are recovered in the summary. We use snowNLP[1] for Chinese word segmentation. We use a manually built set of marks, including question marks and commas for sentence segmentation.

4.1   Tag-Driven Summarization

We conduct extensive experiments to study the performance of CRP model in tag-driven summarizations. We use the DUC, Reuters and Zhihu data sets.

---

[1] https://github.com/isnowfy/snownlp.

Most documents in Reuters and Zhihu data sets are multi-tagged. Only a few documents in DUC data set are multi-tagged. As we are stimulating a scenario of multi-tagged documents, we generate the corpus as follows. Step 1: for a random tag, we select 4 other tags that co-occurred with the given one. Step 2: we gather documents that are tagged by at least two, at most 5 tags generated by step 1. This process may result in a smaller corpus, especially for the DUC data set. We use the gold-standard summary for the DUC data set as the ground truth. We manually generate summaries for Reuter and Zhihu data set. The evaluation metric is ROUGE. We repeat the corpus construction and experiments on each data set for 10 times, and report the average ROUGE result.

As CRP is in its essence a nonparametric probabilistic model, we only compare it with nonparametric models, including (1) GSDMM [16]: a LDA-type model that shrinks the number of topics; (2) LP: a linear programming implementation of the dominant set summarization method [12]; (3) Xmeans: An expansion of K-means for automatically determining the optimal number of clusters [10]; (4) AP [5]: a self-adjusted clustering approach by affinity propagation.

We utilize the following three ranking mechanisms to select sentences. (1) Max: order the sentences by its length, and choose the one with maximal length; (2) LexRank [4]: graphical ranking by structural centrality; (3) DSDR [7]: summarization by matrix reconstruction.

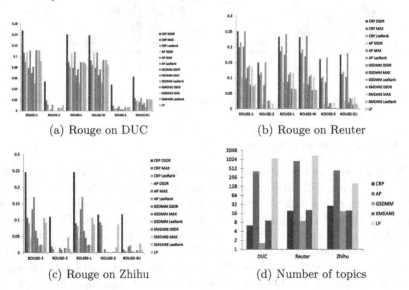

(a) Rouge on DUC

(b) Rouge on Reuter

(c) Rouge on Zhihu

(d) Number of topics

Fig. 2. Performance of tag-driven summarization on three data sets

We have the following observations from Fig. 2. (1) Performances of the CRP models are good and stable. CRP-DSDR achieves the best result in terms of all ROUGE metrics on both DUC and Zhihu data sets. Performance of CRP-DSDR is the second best on the Reuters data set, and it is comparable to that of the best method, in which case, Xmeans-LexRank. (2) For the three ranking strategies,

DSDR is the best for CRP model, on almost every evaluation metric and data sets. But a simple strategy as ordering by length, can also produce acceptable results. (3) As we take a further exploration of the clustering generated by each method, we find that CRP is capable to generate an appropriate number of topics, while apparently LP and Xmeans yield too many topics that need to be cut off, and GSDMM creates too few topics.

4.2 Comparative Summarization

We next study the performance of CRP model in comparative summarizations. We use the Healthcare and Reuters sets. Since we haven't find a summarization system to deal with comparative tag summarizations, our main focus here is to study how the rankings affect the summarization performance. The evaluation metric is ROUGE. As shown in Fig. 3(a) and (b), DSDR performs best when combined with our CRP model. This observation is in agreement with results of tag-driven summarizations.

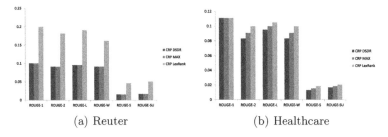

(a) Reuter   (b) Healthcare

Fig. 3. Rouge performance of comparative summarization on two data sets

4.3 Update Summarization

We illustrate an example summary we obtained for the answers of a Zhihu query[2]. There are 372 answers in the query. We divide the answer sets into two parts in chronical order, the first part consists of 272 segments, the remaining part consists of 100 segments published after the fist part. The left column of Fig. 4(a) is the summary in the old data. The right column is the update summary for the new data (the combination of the two parts). We can see that our incremental algorithm successfully updates the summary by recognizing the new topics (topic 34,35). Our algorithm also catches the topic drift. As more people hold the standpoint 13, we replace the old summary with a more rigorous and polite sentence. We further study the changes of popularity of topics. Figure 4(b) demonstrates the size of topics of the 15 most popular topics, in the old and new corpus. As we've seen in Fig. 4(a), the 4 most popular topics are covered in our memory. The topic with significant reshapes, such as topic 13, indicates that the summary for this topic might need a refresh. Again, in our

---

[2] http://www.zhihu.com/question/26472875.

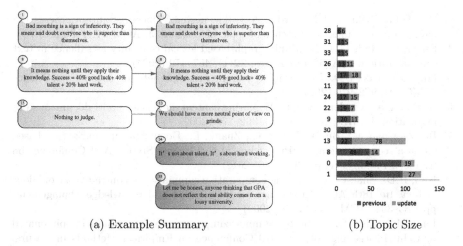

(a) Example Summary   (b) Topic Size

Fig. 4. A case study of the update summary drawn from the Zhihu answers on query "What's your opinion about the grinds in Tsinghua University? Why do people smear/praise them?"

summary in Fig. 4(a), the change is captured. Another interesting discovery is about the background topic. Background topic is the second largest topic. Normally, it will hurt the performance of a summarization system, as it contains a large amount of contents. However, as our CRP model detects the background topic, noisy information is filtered. We also realize the pattern of "popular topic gets more popular", which verifies our assumption in using a Chinese Restaurant Process.

5   Conclusion

In this paper, we present a nonparametric model, based on Dirichlet Process. The model assumes that for each unit text segment, one and only one topic is expressed, which is either a background topic, or a topic drawn from one of the tags that labels the document. The model is capable to detect new topics. We present three straight-forward applications of the model, in tag-driven summarization, comparative summarization and update summarization. We conduct experiments on real world data sets, and present satisfying experimental results. In the future, we will extend the model to opinion mining, and other promising fields.

References

1. Arora, R., Ravindran, B.: Latent Dirichlet allocation and singular value decomposition based multi-document summarization. In: ICDM, pp. 713–718 (2008)
2. Blei, D.V., Ng, A.Y., Jordan, M.I.: Latent Dirichlet allocation. J. Mach. Learn. Res. **3**, 993–1022 (2003)

3. Blei, D.M., Griffiths, T.L., Jordan, M.I.: The nested Chinese restaurant process and bayesian nonparametric inference of topic hierarchies. J. ACM **57**, 1–30 (2010)
4. Erkan, G., Radev, D.: LexRank: graph-based lexical centrality as salience in text summarization. J. Artif. Intell. Res. **22**(1), 457–479 (2004)
5. Frey, B.J., Dueck, D.: Clustering by passing messages between data points. Science **315**(5814), 972–976 (2007)
6. Haghighi, A., Vanderwende, L.: Exploring content models for multi-document summarization. In: HLT-NAACL, pp. 362–370 (2009)
7. He, Z., Chen, C., Bu, J., Wang, C., Zhang, L.: Document summarization based on data reconstruction. In: Proceeding of the Twenty-Sixth AAAI Conference on Artificial Intelligence, pp. 620–626 (2012)
8. Kim, H.D., Zhai, C.: Generating comparative summaries of contradictory opinions in text. In: 18th ACM Conference on Information and Knowledge Management, pp. 385–394. ACM, New York (2009)
9. Paul, M.J., Zhai, C., Girju, R.: Summarizing contrastive viewpoints in opinionated text. In: Proceedings of the 2010 Conference on Empirical Methods in Natural Language Processing, pp. 66–76 (2010)
10. Pelleg, D., Moore, A.: X-means: extending K-means with efficient estimation of the number of clusters. In: Proceedings of the 17th International Conference of Machine Learning, pp. 727–734. Morgan Kaufmann, San Francisco (2000)
11. Ramage, D., Hall, D., Nallapati, R., Manning, C.D.: Labeled LDA: a supervised topic model for credit attribution in multi-labeled corpora. In: Proceedings of the 2009 Conference on Empirical Methods in Natural Language Processing, pp. 248–256. Association for Computational Linguistics, Singapore (2009)
12. Shen, C., Li, T.: Multi-document summarization via the minimum dominating set. In: Proceedings 23rd International Conference on Computational Linguistics, pp. 984–992 (2010)
13. Tang, J., Yao, L., Chen, D.: Multi-topic based query-oriented summarization. In: Proceedings of the Ninth SIAM International Conference on Data Mining, Nevada, USA, pp. 1148–1159 (2009)
14. Teh, Y., Jordan, M., Beal, M., Blei, D.: Hierarchical Dirichlet processes. J. Am. Stat. Assoc. **101**(476), 1566–1581 (2006)
15. Wang, D., Li, T., Zhu, S., Ding, C.: Multi-document summarization via sentence-level semantic analysis and symmetric matrix factorization. In: Proceedings of the 31st Annual International ACM SIGIR Conference on Research and Development in Information Retrieval, pp. 307–314 (2008)
16. Yin, J., Wang, J.: A Dirichlet multinomial mixture model-based approach for short text clustering. In: SIGKDD, pp. 233–242 (2014)

Approaches to Detect Micro-Blog User Interest Communities Through the Integration of Explicit User Relationship and Implicit Topic Relations

Yu Qin[1,2], Zhengtao Yu[1,2(⊠)], Yanbing Wang[1,2], Shengxiang Gao[1,2], and Linbin Shi[1,2]

[1] Institute of Information Engineering and Automation,
Kunming University of Science and Technology, Kunming 650500, China
[2] Key Laboratory of Intelligent Information Processing,
Kunming University of Science and Technology, Kunming 650500, China
ztyu@hotmail.com

Abstract. In order to utilize effectively explicit user relationship and implicit topic relations for the detection of micro-blog user interest communities, a micro-blog user interest community detection approach is proposed. First, we analyze the follow relationship between the users to construct the user follow-ship network. Second, we construct the user interest feature vectors based on the concept of feature mapping to build a user-tag based interest relationship network. Third, we propose to build a guided user interest topic model and construct a topic-based interest relationship network. Finally, we integrate the above-mentioned three kinds of relationship network to construct a micro-blog user interest relationship network. Meanwhile, we propose a micro-blog user interest community detection algorithm based on the contribution of the neighboring nodes. The experiment result turns out that good effect has been achieved through our approach.

Keywords: Feature mapping · Implicit topic · A guided topic model · Contribution of the neighboring nodes

1 Introduction

Regarding the web community detection, scholars have made numerous researches in recent years. Sun and Lin 2013 proposed a probabilistic generative model to detect latent topical communities among users through the capture of user tagging behavior and interest. W. Fan and K. H. Yeung 2014 discussed how profile information could be used to improve community detection in online social networks. Ruan et al. 2013 proposed an approach of combining content with link information in graph structures to detect communities. Li and Pang 2014 proposed a vertices similarity probability (VSP) model to find community structure without the priori knowledge of the type of complex network structure. Wu and

© Springer Science+Business Media Singapore 2015
X. Zhang et al. (Eds.): SMP 2015, CCIS 568, pp. 95–106, 2015.
DOI: 10.1007/978-981-10-0080-5_9

Zou 2014 proposed an incremental community detection method for social tagging systems based on locality-sensitive hashing. Xin et al. 2015 proposed a clustering algorithm for community detection based on the link-field-topic (LFT) model to solve the issue of presetting the number of communities. Good effects have been achieved through all of the above-mentioned approaches. This paper construct a micro-blog user interest relationship network and propose a micro-blog user interest community detection method.

2 Explicit User Relationship

2.1 Construct the User Follow Relationship Network

Through the analysis of follow relationship between the micro-blog users, three types of follow relationship are defined with the details shown in Table 1:

Table 1. Follow relationship between the micro-blog users

Follow relationship definition	Description	Relationship strength(Se)
Follow each other	A follows B and B follows A	1
Unidirectional	A follows B or B follows A	0.7
Strangers	They do not follow each other	0.3

In the table, the relationship strength corresponding to each type of follow relationship declines in turn and the strength depends on a series of tests. The user follow relationship network is indicated in Fig. 1:

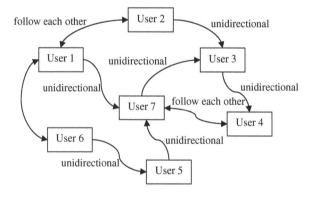

Fig. 1. Micro-blog user follow relationship network

2.2 Construct the User Tag-Based Interest Relationship Network

Feature Selection. We remove all of the personalized tags attached with special symbols or containing English words to obtain such a tag set containing all of the user tags that meet the requirements. Then the frequency of the occurrence of all tags will be calculated, which will also be ranked in descending order to choose those tags ranking in the top as the feature dimension of the user interest feature vectors by setting a threshold.

Feature Mapping-Based Representation of User Feature Vectors. In the process of feature mapping, we utilize the ICTCLAS word segmentation system to segment the words on the long tags so that such tags can be represented by a word set to calculate the average semantic similarity. Assume that the number of the tags for every user is represented by X and the frequency of the occurrence of a tag is represented by x, then f_{ul}, the initial characteristic value of every tag can be expressed by the computational formula (1):

$$f_{ul} = \frac{x}{X} \tag{1}$$

Use $Sl(l_u, l_d)$ to represent the semantic similarity between the user tag and the tag of the feature dimension, where l_u represents the user tag and l_d refers to the tag of the feature dimension. The computational formula can be indicated as (2):

$$Sl(l_u, l_d) = \frac{\sum_{i=1}^{m} \sum_{j=1}^{n} Sim(wu_i, wd_j)}{m \times n} \tag{2}$$

By computing successively the semantic similarity between all of the tags for a user and the tags of the feature dimension, we choose such a user tag that is most similar to the tag of the feature dimension and then multiplies the characteristic value of this tag by the maximum similarity to obtain the computation result. The computational formula for the characteristic value of every feature dimension is indicated as (3) in a feature mapping process:

$$T(l_d) = f_{ul}((l_u)_a) \cdot \max\{Sl((l_u)_a, l_d)\}, (a = 1, 2, 3, \ldots, X). \tag{3}$$

The User Tag-Based Interest Relationship Network. The cosine similarity calculation method can be used to compute the feature vector similarity between the users to represent the strength of interest relationship. We construct a user tag-based interest relationship network, which is indicated in Fig. 2 with the thickness of the edge representing the degree of the interest similarity between the users.

3 Implicit Topic Relations

3.1 Construct the Topic-Based Interest Relationship Network

Guided LDA-Based Micro-Blog User Interest Modeling. We propose a guided LDA model to extract user interest topic. The Bayesian network diagram of this model is shown in Fig. 3:

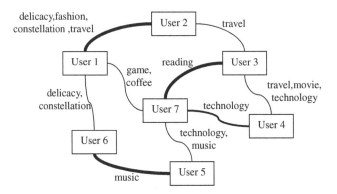

Fig. 2. The user tag-based interest relationship network

Where α, α_c, α_{rt} and α_{re} are respectively the hyper-parameters of θ_d, θ_c, θ_{rt} and θ_{re} that represent separately the topic distribution of the original micro-blog, the commented micro-blog, the reposted micro-blog and the commented micro-blog that has been replied to. γ is the parameter for the extraction of χ distribution, while χ is the influence distribution of the relevant micro-blog sampled according to the Dirichlet distribution of Parameter γ that is created for all of the relevant commented micro-blogs. r is the influential micro-blog extracted from the χ distribution. The type of the micro-blog will be determined according to the data source and the data form when a piece of micro-blog is generated:

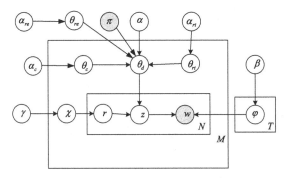

Fig. 3. The Bayesian network diagram

(1) If the micro-blog is original without any comment, then set the value of π to be 0, at this point, it's necessary to obtain θ_d, which is the relationship between this micro-blog and the various topics through the sampling from the Dirichlet distribution of Parameter α.

(2) If it's a user commented micro-blog, then set the value of π to be 1, at this point, it's necessary to obtain θ_c, which is the relationship between the micro-blog that has been commented on this micro-blog and the various topics from

the Dirichlet distribution of Parameter α_c through sampling, and assign the value of θ_c to θ_d, which is the relationship between the micro-blog d and the various topics.

(3) If it's a reposted micro-blog, then set the value of π to be 2, at this point, it's necessary to obtain θ_{rt}, which is the relationship between the micro-blog that has been commented on this micro-blog and the various topics from the Dirichlet distribution of Parameter α_{rt} through sampling, and assign the value of θ_{rt} to θ_d, which is the relationship between the micro-blog d and the various topics.

(4) If it's a reply to the other user's comments on a micro-blog, then set the value of π to be 3. In this case, the weight of influence from the comment that is replied to and the original micro-blog that has been commented on the topic distribution of this reply is different. When μ, the influencing parameter is introduced to represent the influencing weight of the commented micro-blog that is replied to, the influencing weight of the original micro-blog that has been commented is then $1 - \mu$. At this point, conduct integrated computation on both of the topic distribution of the commented micro-blog that is replied to and the original micro-blog that has been commented to obtain a mixed topic distribution, whose value will be assigned to θ_d, which is the relationship between the micro-blog d and the various topics.

As above, the computation formula for the topic distribution of the mentioned micr -blog types is shown in (4):

$$p(\theta|\alpha, \mu) = \begin{cases} \theta_d & \pi = 0, \alpha = \alpha \\ \theta_c & \pi = 1, \alpha = \alpha_c \\ \theta_{rt} & \pi = 2, \alpha = \alpha_{rt} \\ \mu\theta_{re} + (1-\mu)\theta_c & \pi = 3, \alpha = \{\alpha_{re}, \alpha_c\} \end{cases} \tag{4}$$

For any of the original micro-blog that has been commented by the other users, since the other users' comments will influence to some extent the topic distribution of this micro-blog, we shall include this original micro-blog and its comments in a data set S_d, and then add a χ_d distribution that is obtained through the sampling of the Dirichlet distribution of Parameter γ to this data set. For every word contained in the micro-blog text, first extract an influential micro-blog r, which is included in the data set S_d constituted by the original micro-blog and its comments, from the χ_d distribution. Then obtain θ_γ which is the relationship between this influential micro-blog and the various topics through the sampling of the Dirichlet distribution of Parameter γ_d to extract the topic z based on such a relationship. After that, extract a word from φ to fill out the corresponding space on the micro-blog. In this model, the computation formula for the probability distribution of θ is shown in (5):

$$P(\theta|\alpha, \mu, \gamma_d) = xP(\theta|\alpha, \mu) + (1 - x)P(\theta_\gamma|\gamma_d) \tag{5}$$

Where x is the Boolean value, which is set to 1 when it's a reposted micro-blog, a comment, an original micro-blog without any comment from the other users or a reply to the comment. Otherwise x should be set to 0, representing that this micro-blog is original.

Model Deduction and the Extraction of User Interest Topic. The joint probability distribution of micro-blog, words and topic can be expressed formally by Formula (6):

$$P(r, z, w | \varphi, \theta, \chi) = \prod_{d \in D} \frac{\Delta(M_d^r + \gamma_d)}{\Delta(\gamma_d)} \cdot \prod_{d \in D} \frac{\Delta(N_d^z + \alpha)}{\Delta(\alpha)} \cdot \prod_{z \in T} \frac{\Delta(N_z^w + \beta)}{\Delta(\beta)} \quad (6)$$

Decompose Formula (6) and iterate the Gibbs sampling according to Formula (7):

$$P(z_i = z', r_i = r' | z_{-i}, r_{-i}, w) = \frac{P(z, r, w)}{P(z_{-i}, r_{-i}, w_{-i})} \cdot \frac{1}{P(w_i | z_{-i}, r_{-i}, w_{-i})} \quad (7)$$

The obtained posterior distribution formula can be indicated by Formula (8):

$$P(z_i = z', r_i = r' | z_{-i}, r_{-i}, w) \propto \frac{P(z, r, w)}{P(z_{-i}, r_{-i}, w_{-i})}$$

$$= \frac{N_{r'z'}^{-i} + \alpha}{N_{r'}^{-i} + T\alpha} \cdot \frac{M_{dr'}^{-i} + \gamma_d(r')}{\sum_{r \in S_d} (M_{dr'}^{-i} + \gamma_d(r))} \cdot \frac{N_{z'w_i}^{-i} + \beta}{N_{z'}^{-i} + V\beta} \quad (8)$$

Since the sampling of words and topics meets the requirement of multinomial distribution, the results of θ_d, θ_{rt}, θ_c, θ_{re}, φ_z and χ can be expressed separately by Formulas (9), (10), (11), (12), (13) and (14) as indicated below:

$$\theta_d = \frac{N_{dz} + \alpha}{N_d + T\alpha} \quad (9)$$

$$\theta_c = \frac{N_{cz} + \alpha_c}{N_c + T\alpha_c} \quad (10)$$

$$\theta_{rt} = \frac{N_{rtz} + \alpha_{rt}}{N_{rt} + T\alpha_{rt}} \quad (11)$$

$$\theta_{re} = \frac{N_{rez} + \alpha_{re}}{N_{re} + T\alpha_{re}} \quad (12)$$

$$\varphi_z = \frac{N_{zw} + \beta}{N_z + V\beta} \quad (13)$$

$$\chi_d(r) = \frac{M_{dr} + \gamma_d(r)}{\sum_{r \in S_d} (M_{dr} + \gamma_d(r))} \quad (14)$$

For every user, the probability distribution of the user under each topic can be obtained to get the user-topic feature vectors when the probabilities under each topic have been summed up and the result has been divided by the number of the user's micro-blogs. Also for every topic, the user-topic probability can be calculated through Formula (15):

$$P_u(z_i) = \frac{\sum_{i=1}^{N} P(z_i)}{N}. \quad (15)$$

The Topic-Based Interest Relationship Network. Through the guided LDA-based method to extract the topic of user interests, the interest similarity between the users can be computed. We construct a micro-blog topic-based interest relationship network that is shown in Fig. 4 with the thickness of the edge representing the degree of the interest similarity between the users.

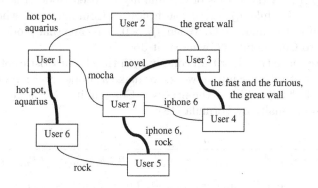

Fig. 4. The topic-based interest relationship network

4   Approaches to Detect Micro-Blog User Interest Communities Through the Integration of Explicit User Relationship and Implicit Topic Relations

4.1   Network Convergence

Use S_e, whose value is defined in Table 1 to represent the strength of interest relationship in the user follow-ship network. The total strength of the interest relationship between the users can be defined formally through Formula (16) as below:

$$S_t = S_e(\lambda_{hL}S_{hL} + \lambda_{hM}S_{hM}) \tag{16}$$

Where λ_{hL} is the influential parameter for the strength of interest relationship based on user tag, while λ_{hM} is the influential parameter for the strength of interest relationship based on micro-blog topic. Through the analysis of the micro-blog tag and the content, we set separately the values of both parameters to be $\lambda_{hL} = 0.3$ and $\lambda_{hM} = 0.7$ to calculate S_t, the total strength of interest relationship between the users. Meanwhile by setting a threshold value, we delete the edges with the total strength of interest relationship lower than the threshold to construct finally a user interest relationship network.

4.2   Approaches to Detect Micro-Blog User Interest Communities

Node Contribution. Node contribution represents the degree of the contribution made by the nodes to the community. The greater the node contribution is, the more necessary the nodes will be covered in the community. The node contribution can be represented formally by Formula (17):

$$f_G^A = f_{G+\{A\}} - f_{G-\{A\}} \tag{17}$$

Where $f_{G+\{A\}}$ is the fitness of Community G when Node A is subordinate to Community G, while $f_{G-\{A\}}$ is the fitness of Community G when Node A is not covered in Community G. When $f_G^A > 0$, it means that Node A makes a contribution to Community G, whereas it doesn't.

In order to make the algorithm comply with the characteristics of the microblog user interest relationship network, we make a proper revision on f_G (Lancichinetti et al. 2009), which is the fitness of Community G. The revised f_G can be represented formally by Formula (18):

$$f_G = \frac{m * k_{d\_in}^G + n * k_{s\_in}^G}{(m * k_{d\_in}^G + n * k_{s\_in}^G + m * k_{d\_out}^G + n * k_{out\_in}^G + t * k_{in\_out}^G)^\alpha} \tag{18}$$

We set the value of Parameter α to be 1. The parameters of m, n and t are the contribution coefficients introduced for the different subordinative tendencies of the community, where m represents the contribution coefficient of the nodes that are directed to each other, n refers to the contribution coefficient of the nodes that direct to the nodes within the community and t represents the contribution coefficient of the nodes that are within the community but direct to the nodes of the other communities. In order to normalize the contribution coefficients, we define the contribution coefficient vector as $\{\beta_1^i, \beta_2^i, \beta_3^i..., \beta_k^i\}$, where β_k^i is the degree of contribution made by Node i, when it's subordinate to Community G. It conforms to the requirement that the sum of the coefficients for the contributions made by any node in Community G to Community G is equal to 1. And it can be represented formally by Formula (19):

$$\sum_{k=1}^{n} \beta_k^i = 1, 0 \le \beta \le 1 \tag{19}$$

Where i is an arbitrary node in the network and n is the number of the contribution coefficients.

Community Overlapping Degree. Community overlapping degree represents the overlapping degree between the communities. The greater the community overlapping degree is, the larger the overlapping area between the communities will be. The community overlapping degree can be expressed formally by Formula (20):

$$overlap(G_i, G_j) = \frac{|G_i \cap G_j|}{|\min(G_i, G_j)|} \tag{20}$$

where $\min(G_i, G_j)$ represents the number of the nodes in Community G_i or Community G_j, whichever has the minimum nodes.

The Algorithm Process.

(1) Calculate P, which is the value of all nodes in the network and then store successively the nodes, which are ranked in descending order according to the value of P into Queue Q.

(2) De-queue the nodes at the top from Queue Q and take the front node N as the initial community G.

(3) Calculate f_G^A, which is the node contribution made by all of the neighboring nodes in Community G. If the maximum value of the node contribution made by any of the neighboring node is $\max(f_G^A) > 0$, then include this node with the greatest contribution into Community G and remove it from Queue Q. Repeat this process and if $\max(f_G^A) < 0$, which means that the node contribution of every neighboring node is less than 0, it indicates that Community G is saturated and all of the members in Community G have been detected. Then skip to Step (4) to detect other communities in the network.

(4) If Queue Q is not empty, then skip to Step (2). However if Queue Q is empty, it means that all of the communities in the network have been detected. At this point, skip ahead to Step (5).

(5) Compute the overlapping degree of all detected communities in pairs and merge those communities whose overlapping degree is higher than the threshold value.

5 Experiments

5.1 Experimental Data Set

We select manually totally 1000 celebrity users who have exerted great influence on Sina micro-blog in addition to the crawling of their micro-blog information, including their follower list, the user's custom tag, the tag of their followees, and the posts published by the user on the micro-blog. The basic information about the crawled experimental data set is shown in Table 2.

Table 2. Sina micro-blog data set

The number of users	The number of user tags	The number of micro-blog
1000	8947	50000

5.2 Evaluation Index for Community Division

Sales-Pardo et al. 2007 proposed EQ, the expanded modularity function and defined it formally by Formula (21):

$$EQ = \frac{1}{2m} \sum_{ij} \frac{1}{o_i o_j} (A_{ij} - \frac{k_i k_j}{2m}) \delta(c_i c_j) \tag{21}$$

5.3   Experimental Settings and Result Analysis

Experiment 1: Contrast Experiment on the Interest Community Division Modularity Based on Different Algorithms

We choose to conduct contrast experiment through CPM (Clique Percolation Method) (Palla et al. 2005), CORPA label propagation algorithm (Raghavan et al. 2007) and our approach, where the parameters in Formula (18) are set to be $m = 0.5$, $n = t = 0.25$ and $\alpha = 1$ in the experimental process. The community detection results are shown in Fig. 5:

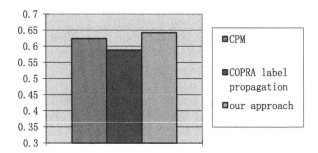

Fig. 5. Contrast experiment on the interest community division modularity based on different algorithms

Figure 5 indicates that our approach is slightly superior to the CPM and CORPA label propagation algorithm from the perspective of community division modularity. Such an experiment result can be explained easily in theory: the CPM has to detect constantly the maximum sub-graph consisting of k-Cliques that are connected mutually in the network during the community division. Therefore this algorithm must be defined strictly. The CORPA label propagation algorithm will determine which community the nodes are subordinate to according to the number of the tags brought by the neighborhood set. It doesn't consider such a situation that different neighboring nodes might give a different contribution to the community division, which also means that the connection between the members in the community is much closer than the connection between the communities.

Experiment 2: Influence of Explicit User Relationship and Implicit Topic Relations on the Division of Interest Communities

In order to verify the influence of the explicit user relationship and implicit topic relations on the division of interest communities, we've designed the contrast experiment. The results of the contrast experiment in three phases are shown in Fig. 6:

The experiment result shows that better effect will be achieved in the division of micro-blog user interest communities when both of the explicit user relationship and implicit topic relations are taken into account.

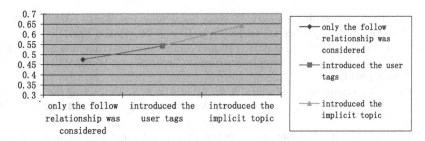

Fig. 6. Influence of explicit user relationship and implicit topic relations on the division of interest communities

6   Conclusions

This paper analyzes the mutual follow relationship between the users in a micro-blogging environment, and constructs a user tag-based user interest model according to the concept of feature mapping. Also by taking the user behavior as the supervisory information, we build a guided LDA based topic extraction model to extract the micro-blog topics so as to construct a micro-blog user interest relationship network finally. Regarding the community detection algorithm, we propose a micro-blog user interest community detection algorithm based on the contribution of the neighboring nodes by combining the concept of local information with the characteristics of the micro-blog user interest relationship network. The experiment result proves the effectiveness of our approach in the task about the detection of the micro-blog user interest communities.

Acknowledge. This paper is supported by the China National Nature Science Foundation (No.61175068, 61472168, 61163004), and The Key Project of Yunnan Nature Science Foundation (No.2013FA130). Corresponding author is Zhengtao Yu, E-mail ztyu@hotmail.com.

References

Fan, W., Yeung, K.H.: Incorporating profile information in community detection for online social networks. Phys. A **405**, 226–234 (2014)

Liu, Q., Li, S.: Word similarity computing based on how-net. Int. J. Comput. Linguist. Chin. Lang. Process. **7**, 59–76 (2002)

Lancichinetti, A., Fortunato, S., Kertesz, J.: Detecting the overlapping and hierarchical community structure of complex networks. New J. Phys. **11**, 033015 (2009)

Li, K., Pang, Y.: A unified community detection algorithm in complex network. Neurocomputing **130**, 36–43 (2014)

Newman, M.E.J., Girvan, M.: Finding and evaluating community structure in networks. Phys. Rev. E **69**, 026113 (2004)

Palla, G., Derényi, I., Farkas, I., Vicsek, T.: Uncovering the overlapping community structure of complex networks in nature and society. Nature **435**, 814–818 (2005)

Raghavan, U.N., Albert, R., Kumara, S.: Near linear time algorithm to delect community structures in large-scale networks. Phys. Rev. E **76**, 036106 (2007)

Ruan, Y., Fuhry, D., Parthasarathy, S.: Efficient community detection in large networks using content and links. In: 22nd International Conference on World Wide Web, pp. 1089–1098 (2013)

Sales-Pardo, M., Guimera, R., Moreira, A.A., Amaral, L.A.N.: Extracting the hierarchical organization of complex systems. Proc. Natl. Acad. Sci. U.S.A. **104**, 15224–15229 (2007)

Sun, X., Lin, H.: Topical community detection from mining user tagging behavior and interests. J. Am. Soc. Inform. Sci. Technol. **64**, 321–333 (2013)

Wu, Z., Zou, M.: An incremental community detection method for social tagging systems using locality-sensitive hashing. Neural Netw. **58**, 14–28 (2014)

Yu, X., Yang, J., Xie, Z.: A semantic overlapping community detection algorithm based on field sampling. Expert Syst. Appl. **42**, 366–375 (2015)

Supervised Link Prediction
Using Random Walks

Yuechang Liu[1,3]([✉]), Hanghang Tong[2], Lei Xie[4], and Yong Tang[1]

[1] South China Normal University, Guangzhou 510631, Guangdong, China
ychangliu@gmail.com, ytang@scnu.edu.cn
[2] School of Computing, Informatics and Decision Systems Engineering,
Arizona State University, Tempe, USA
hanghang.tong@asu.edu
[3] School of Computer Science, Jiaying University,
Meizhou 514015, Guangdong, China
[4] Computer Science, Hunter College, The City University of New York,
New York, USA
lei.xie@hunter.cuny.edu

Abstract. Network structure has become increasingly popular in big-data representation over the last few years. As a result, network based analysis techniques are applied to networks containing millions of nodes. Link prediction helps people to uncover the missing or unknown links between nodes in networks, which is an essential task in network analysis.

Random walk based methods have shown outstanding performance in such task. However, the primary bottleneck for such methods is adapting to networks with different structure and dynamics, and scaling to the network magnitude. Inspired by Random Walk with Restart (RWR), a promising approach for link prediction, this paper proposes a set of path based features and a supervised learning technique, called Supervised Random Walk with Restart (SRWR) to identify missing links. We show that by using these features, a classifier can successfully order the potential links by their closeness to the query node. A new type of heterogeneous network, called Generalized Bi-relation Netowrk (GBN), is defined in this paper, upon which the novel structural features are introduced. Finally experiments are performed on a disease-chemical-gene interaction network, whose result shows SRWR significantly outperforms standard RWR algorithm in terms of the Area Under ROC Curve (AUC) gained and better than or equal to the best algorithms in the field of gene prioritization.

1 Introduction

Link prediction is an important task for analysing social networks which also has applications in other domains like, information retrieval, bioinformatics and e-commerce. Such links between individuals may be missing due to imperfect acquirement processes (e.g. friends in real-world do not form a virtual connection

© Springer Science+Business Media Singapore 2015
X. Zhang et al. (Eds.): SMP 2015, CCIS 568, pp. 107–118, 2015.
DOI: 10.1007/978-981-10-0080-5_10

in online social networks [15].), or unknown due to current knowledge (e.g. the interaction between certain diseases and virulence genes [2]). Link prediction helps people to uncover the missing or unknown links between nodes in networks, while it is also very practical in information recommendation used by real world application like online social networks.

Basically link prediction problem is usually regarded as the generalization of the problem of proximity measurement. In these settings, the data is compiled into a weighted graph representation with nodes representing objects, and edges representing associations between nodes. Then problem is transformed into the computation of node ordering of interested objects by some measure of proximity for a given query node.

One common approach for solving the Link Prediction problem is using supervised learning algorithms, with careful selection of relevant features [10]. Such features are usually derived from network topology like node degrees, common neighbours, shortest paths et al. These features proposed in the literature are to some extent effective in proximity measuring. However, they are far from sufficient when adaptability and scalability are considered.

Another promising method for proximity measurement is based on random walks on graphs, for instance personalized PageRank [3] or Random Walk with Restart [18,20]. These are usually regarded as unsupervised learning approaches. Traditional random walk based similarity measurement only runs a standard iteration procedure or matrix calculation, which regards networks as homogeneous, which usually contradicts intuition because many real world networks are heterogeneous. For example, when one wants to find candidate papers published in certain year y to cite it is more effective to find the papers that are frequently cited in y than all the papers published in y. The intuition can be mapped to path constraint when running the citation-publication year network [13]. Such fact makes traditional random walk based methods hardly can achieve satisfactory performance scalably.

This paper makes contributions to the questions as follows. Firstly, a specific heterogeneous network call Generalized Bi-Relational Network (GBN for short), which distinguishes the links into two categories, namely similarity links and association links, is formalized. Then a series of structural features, which stems from the Random Walk with Restart (RWR) method, are defined and fitted into a supervised learning framework. GBN is capable of representing many kinds of heterogeneous network in real world application (like disease-gene networks and author-citation networks). For the Supervised Random Walk with Restart (called SRWR) framework, the feature design is mainly discussed. At last, experiments are conducted upon a gene-disease association network. Different combinations of the features are run and compared on their Area Under ROC Curve (AUC) gained. The experiment results reveal that our method outperforms traditional RWR algorithm.

In the remainder of this paper, we first review the related work in depth. Then the detailed formalization and description of our framework and data set are described next. Subsequently we focus on the experiment design, result display and analysis. At last the paper is concluded.

2   Related Work

The algorithms solving link prediction problem can be roughly categorized into two classes: unsupervised and supervised learning based, both of which are based on the computation of node proximity[1]. Several measurements of node proximity have proposed in the literature. Just name a few, there are Common-Neighbours (CN), Cosine Similarity (CS) [19], Jaccard Coefficient (JC) [11], Katz Measure (KM) [12], Shortest Path (SP), Friends Measurement [8], etc. More thorough survey and comparison of these measurements can be found in several review papers [5,10,16].

For the first category, random walk has been a very famous spectrum of unsupervised learning algorithms for its simplicity and elegance in measuring proximity of networks. Among so many of its descendants, personal PageRank should be the most famous one [17]. Another variants - Random Walk with Restart (RWR) has also attracted great attention of researchers in the last decades [18,20]. Though RWR performs well in many real world problems, its unsupervisedness nature makes it hardly always adaptive well to various domains.

The second category has been attracting intensive research in recent years. Supervised learning based link prediction algorithms was first introduced by Liben-Nowell and Kleinberg in 2003 [15], who studied the usefulness of graph topological features by testing them on bibliographic data sets. Hasan et al. extended their work in 2006 [9] by identifying a list of easily-computable features (pacifically, for the coauthorship domain.) and compared different models (namely decision tree, kNN, multilayer perception, SVM, RBF network) on the features and data set. More recently, W.J. Cukierski et al. incorporated 94 distinct graph features in their random Forests classifier and achieved impressive results in predicting links on Flickr dataset [6]. Lars and Jure also proposed a supervised random walk algorithm for link prediction and recommendation especially for social networks [1]. In their algorithm, they focused on learning weights for the edges in social network, on the basis of rich attribute information of the nodes (e.g. users of Facebook), then transferred the problem of classification to optimization. Though it performs well on user recommendation in Facebook network, Lars and Jure's method requires rich information of user node features like age, gender and hometown, and edge features like interaction activity. And what's worse is that their method is computation intensive to learn all of the weights in the matrix, which makes their methods harder to be practical.

Random walk based algorithms for link prediction is also improved by more careful feature selections and supervised learning. Along this thought Ni Lao and Cohen proposed Path-Constrained Random Walks (PCRW) framework. By the investigation of a paper-publication network, they found that only subset of paths between nodes are relevant [13]. Other related work are those studying link prediction in heterogeneous networks. Traditional random walk based algorithms are proposed for homogeneous networks (that do not make distinction

---

[1] Also named "similarity", "closeness" or other similar words in literature, they will be used interchangable in following text.

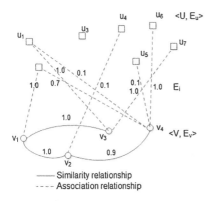

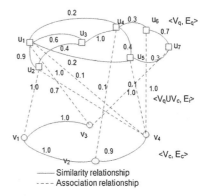

Fig. 1. Bi-relation Network in [21]

Fig. 2. Generalized Bi-relation Network

on the different types of nodes and links). In 2009 Jing Xia proposed a kind of heterogeneous network: bi-relation network and shows its application in the modeling or co-author network. When fed into the RWR algorithm framework, their IAD-based RWR achieves higher efficiency compared to the traditional power method based RWR [21]. The heterogeneousness has to be taken into account to achieve better link prediction effectiveness in some real life domains [23].

3   Supervised Random Walk with Restart

3.1   Generalized Bi-Relational Network

In traditional RWR framework, the proximity scores of candidate nodes to a query node is calculated according to the formula [20]:

$$q_i = c\tilde{W}q_i + (1-c)e_i \tag{1}$$

Equation (1) formulates the calculation of proximity scores of candidate nodes with respect to a given query node i. Generally RWR runs on general networks that make no difference to all of the nodes or edges. Such general network is called Single-Relational Networks (SRN) in this paper. An SRN is formally defined to be a network $N = (V, E, W)$ where V is the set of nodes, E for the edges and W is the function that assigns each edge with a real valued weight. For the clarification of our method, we formally define a new network model, which is called Generalized Bi-relational Network (GBN), as follows.

Definition 1 (Generalized Bi-relational Network). *A Generalized Bi-relation Network (GBN) is a network* $< V_q \cup V_c, E_q \cup E_c \cup E_i, W >$, where: V_q and V_c are respectively the set of query nodes and candidate nodes, and E_q, E_c and E_i are the set of edges within V_q, within V_c and inter-connect V_q and V_c; W has the same definition as in GBN. Particularly, edges in E_q and E_c are called **similarity** edges and **association** edges, respectively.

Similar network structure is also defined in the literature. In [21], the author proposes a type of bi-relational network, where the relations are defined to be two disjoint sets and similarity relations only happen between different node set. GBN allows similarity relations within any node set (illustrated in Fig. 2). It is obvious that the definition of GBN in this paper is more general than the one in Fig. 1. Indeed, the BRN proposed in [21] can easily be represented by our model, but not vice versa.

Given a GBN $< V_q \cup V_c, E_q \cup E_c \cup E_i, W >$, a **transition matrix** is defined to be:

$$W_{N \times N} = \begin{bmatrix} M_q & M_{qc} \\ M_{cq} & M_c \end{bmatrix}, \text{ where } N = |V_q| + |V_c| \text{ and}$$

$M_{q(i,j)} = W(< i, j >) \text{ s.t. } < i, j > \in E_q,$
$M_{c(i,j)} = W(< i, j >) \text{ s.t. } < i, j > \in E_c,$
$M_{qc(i,j)} = W(< i, j >) \text{ s.t. } < i, j > \in E_i,$
$M_{cq} = M_{qc}^T.$

3.2 Our Method

Traditionally, RWR is solved using iteration [18], or matrix computation [20]. Matrix computation solution of RWR consists in matrix inversion and multiplication, which is not practical for large network at all. Though researchers have found alternative solutions under matrix computation for RWR [20], they usually rely on heave-load pre-computation based on matrix refraction. However, it is still a time-consuming task to do the matrix refraction (SVD, LU, etc.) when the network is overly large. It is even more impossible when the network keeps changing dynamically, or/and the pre-computation needed is excessively too much. Just think about the scenario when using RWR pre-computation for user recommendation in a social network the server has to do a matrix refraction for every user node! Once the network is changed, the computation has to be invoked from scratch. Iterative RWR algorithm is still a practical choice for many applications.

To improve the efficiency of traditional RWR, supervised learning gives a practical route.

$$q_i = c\tilde{W}q_i + (1-c)e_i$$
$$= \lim_{n \to \infty} (c^n \tilde{W}^n q_i + \sum_{i=0}^{n}(1-c)c^i \tilde{W}^i e_i)$$
$$= \lim_{n \to \infty} (\sum_{i=0}^{n}(1-c)c^i \tilde{W}^i e_i)$$
$$= (1-c)\sum_{i=0}^{\infty} c^i \tilde{W}^i e_i \tag{2}$$

From simple inference one can find that the RWR algorithm is mathematical limit of a linearly combination of some terms (see Eq. 2), which gives us inspiration that for $\tilde{W}^i$s should be important features that have impact on the final proximity scores. Indeed, $\tilde{W}^i[u, v]$ semantically represents the probability that the particle reaches node v from node u along the paths having length of i.

Problem 1. Given a GBN $\langle V_q \cup V_c, E_q \cup E_c \cup E_i, W \rangle$ and a node $s \in V_q$, we aim to compute a function $score(s, v)$ for every $v \in V_c$ that assigns a score value to each candidate node. Additionally, we assume that a training set $D = \{\langle \boldsymbol{x}_i, y_i \rangle | i = 1, 2, \cdots, n\}$ exists, where $\boldsymbol{x}_i = \langle \tilde{W}^2[u, v], \tilde{W}^3[u, v], \cdots, \tilde{W}^{N+1}[u, v] \rangle$, $u \in V_q$ and $v \in V_c$, and $y_i \in \{0, 1\}$.

Note: In the above problem formulation, we use $\langle \tilde{W}^2[u, v], \tilde{W}^3[u, v], \cdots, \tilde{W}^{N+1}[u, v] \rangle$ as the features under the presumption that there is no known direct path between node u and v so that u has to go through a path at least of length of 2 to reach v. The assumption is practical when people wants to find causal genes for totally new diseases or chemicals for novel target genes.

The Optimization Problem. *Under the same GBN network, the training data set D contains the information that the query node will have an association to candidate nodes with $y_i = 1$ or not with $y_i = 0$. So, we aim to set the parameters $\boldsymbol{w}$ for the respective features and finally it will assign scores to order the nodes.*

Thus, like the primal optimization problem definition of Support Vector Machine (SVM) we define the optimization problem to find the optimal set of parameters (mainly $\boldsymbol{w}$) of the features $\langle \tilde{W}^2[u, v], \tilde{W}^3[u, v], \cdots, \tilde{W}^{N+1}[u, v] \rangle$ as follows.

$$\mathcal{L}(\boldsymbol{w}, b, \alpha) = \min_{\boldsymbol{w}, b, \alpha} \frac{1}{2} \parallel \boldsymbol{w} \parallel^2 - \sum_{i=1}^{n} \alpha_i (y_i(\boldsymbol{w}^T \Phi(\boldsymbol{x}_i) + b) - 1) \tag{3}$$

In Eq. 3 $\Phi(x_i)$ is the kernel function. Then, the optimization problem is defined to be:

$$\theta(\boldsymbol{w}) = \max_{\alpha_i \geq 0} \mathcal{L}(\boldsymbol{w}, b, \alpha) \tag{4}$$

Like traditional SVM algorithm, Eq. 4 is solved by first being transformed into its dual representation. Thus the optimization problem can be solved by general SVM package like [4].

3.3 More Alternative Features

As mentioned above, Eq. 3 presents a series of features to use, which consist of the probability on the paths of given length:

$$FP\_features = \{\tilde{W}^2, \tilde{W}^3, \cdots, \tilde{W}^{N+1}\} \tag{5}$$

where $\tilde{W}$ is normalized transition matrix. In this paper we call it Full Path (FP) features. Figure 3 illustrates all of the paths (with passing edges in bold lines) of length 3 between the vertices u_1 and v_1 (i.e. the paths $u_1 \rightarrow u_3 \rightarrow u_2 \dashrightarrow v_1$, $u_1 \rightarrow u_4 \dashrightarrow v_2 \dashrightarrow v_1$, $u_1 \dashrightarrow v_4 \rightarrow v_2 \rightarrow v_1$, $u_1 \dashrightarrow v_4 \dashrightarrow u_2 \dashrightarrow v_1$). When calculating the transition probability from u_1 to v_1 through 3 hops, FP features sums up the values of all of the possibilities.

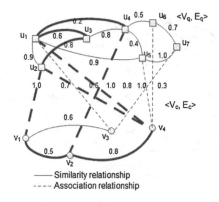

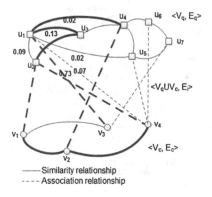

Fig. 3. Full Path (FP) feature

Fig. 4. Full Path for Heterogeneous Network (FPH) feature

Full Path for Heterogeneous Network Features. Take the heterogeneousness of network under consideration, the transition of particles from a node to its adjacent nodes may take different effects according to the edge type. This is accounted by the introduction of a decay factor λ. Indeed, FPH features are basically full path features but different from FP features in that the transition matrix W is decomposed into four parts: $W = \begin{bmatrix} W_q & W_{qc} \\ W_{cq} & W_c \end{bmatrix}$ and the parameter λ serves as decay factor to distinguish whether the random walker jumps via association or similarity links.

$$FPH\_features = \{\tilde{W}^2, \tilde{W}^3, \cdots, \tilde{W}^{N+1}\}, \text{ where } \tilde{W} = \begin{bmatrix} M_q & M_{qc} \\ M_{cq} & M_c \end{bmatrix}, \text{ and}$$

$$M_{q(i,j)} = \begin{cases} W_{q(i,j)}/\sum_j W_{q(i,j)} & \text{if } \sum_j W_{qc(i,j)} = 0 \\ (1-\lambda)W_{q(i,j)}/\sum_j W_{q(i,j)} & otherwise \end{cases}$$

$$M_{c(i,j)} = \begin{cases} W_{c(i,j)}/\sum_j W_{c(i,j)} & \text{if } \sum_j W_{qc(j,i)} = 0 \\ (1-\lambda)W_{c(i,j)}/\sum_j W_{c(i,j)} & otherwise \end{cases}$$

$$M_{qc(i,j)} = \begin{cases} \lambda W_{qc(i,j)}/\sum_j W_{qc(i,j)} & \text{if } \sum_j W_{qc(i,j)} \neq 0 \\ 0 & otherwise \end{cases}$$

$$M_{cq(i,j)} = \begin{cases} \lambda W_{cq(i,j)}/\sum_j W_{cq(i,j)} & \text{if } \sum_j W_{cq(i,j)} \neq 0 \\ 0 & otherwise \end{cases}$$

The same features are used in the literature [14], where the algorithm RWRH is proposed for gene prioritization in PPI networks. RWRH takes FPH features as the preprocessing of the transition matrix and feeds it to a standard RWR solver. Figure 4 illustrates FPH feature value of the node u_1. Compared to FP feature value of the same node in Fig. 3, the amount of paths stays the same but with smaller weights on the related edges after the multiplication of parameter $\lambda = 0.9$.

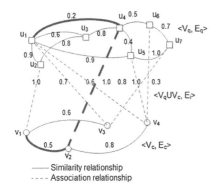

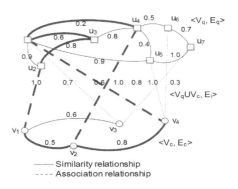

Fig. 5. Balanced Path (BP) feature

Fig. 6. Similarity-Association Bridge (SAB) feature

Balanced Path Features. Balanced Path (BP) features are first used in the Bi-Random Walk (BiRW) algorithm proposed in the paper [22]. To compute BP features, the GBN is decomposed into three parts: two similarity subnetworks and similarity-association subnetwork, which are represented by the transition matrix W_q, W_c and W_{qc} respectively. Then, BP feature set is defined to be:

$$BP\_features = \{W_q W_{qc} W_c, W_q^2 W_{qc} W_c^2, \cdots, W_q^N W_{qc} W_c^N\} \tag{6}$$

Intuitively, BP features only include those paths with equal number of query and candidate nodes. In Fig. 5, one can find that there is only one BP path of length 3 between u_1 and v_1 ($u_1 \rightarrow u_4 \dashrightarrow v_2 \dashrightarrow v_1$).

Similarity-Association Bridge Features. Similarity-Association Bridge (SAB) features can be seen as the generalization of FP and BP features. SAB considers any combinations of the number of the query nodes and candidate nodes on both sides of association edge (like a bridge between the two types of nodes). Formally, SAB features are defined to be:

$$SAB\_features = \{W_q W_{qc} W_c, W_q^2 W_{qc} W_c, \cdots, W_q^i W_{qc} W_c^j, \cdots, W_q^N W_{qc} W_c^N\} \tag{7}$$

As an illustration Fig. 6 depicts all of the possible paths of length 3 from u_1 to v_1 under SAB feature definition: $u_1 \dashrightarrow v_4 \rightarrow v_2 \rightarrow v_1$, $u_1 \rightarrow u_3 \rightarrow u_2 \dashrightarrow v_1$ and $u_1 \rightarrow u_4 \dashrightarrow v_2 \dashrightarrow v_1$.

4   Experiment

4.1   Data Set

We run our experiment on curated chemicalCgene interaction, —chemicalCdisease association, and geneCdisease association network that were retrieved from the

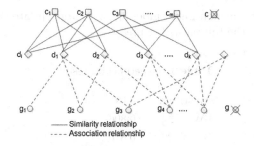

Fig. 7. Chemical-Disease-Gene network topology

Comparative Toxicogenomics Database (CTD) [7]. The topology of the network is illustrated in Fig. 7. In such a network people may be interested in such questions as "which genes are potentially associated with a given disease". There are 3,522 chemicals, 2,845 diseases and 12051 genes in the network. Though semantically the network includes three types of nodes, we transform it into a GBN, by taking chemicals and diseases as the query nodes. For the fairness consideration, during this step some of the nodes are removed, singleton nodes (like the node "c" and "g" labeled with a red cross in Fig. 7). The result on these nodes would only randomly generated and makes no sense.

4.2   Experiment Setting and Result

Our experiment runs a 5-fold cross validation on the given GBN. Each fold of validation consists in 20 % of the diseases which is taken to be test set, another 80 % of the diseases are chosen for the construction of training set. For all of the training disease, randomly select a set of pairs of diseases and genes that label whether there is a link between them. In the experiment we compared standard RWR, two state-of-the-art gene prioritization algorithms RWRH [14] and BiRW [22] with SRWR. In the running of RWRH, the parameter λ is set to be 0.9 which is recommended in [14] AUC value in our experiment. For BiRW, we set $l = r = 4$ which is also recommended in [22]. The same parameters are used in the computation of FPH and BP features. For SRWR, we tested the model of logistics and RBF. The result is displayed in Table 1. From the result one can easily find that SRWR outperforms standard RWR with any model and features, which shows the better scalability of SRWR to specific data set. The best combination of SRWR slightly gets better performance than BiRW (0.886 vs. 0.870), and achieves equal AUC values as RWRH.

For the features proposed in this paper, FPH features achieve the highest AUC, which shows its superiority towards others. The phenomenon coincides with RWRH in gene prioritization domain. RWRH shows high competitiveness compared to other popular algorithms [14]. The four types of features can be roughly ordered as $FPH > FP \approx BP > SAB$ according to the AUC values they get. For the models selected, RBF for SVM is largely superior to logistics model, which tells the fact that non-linear model is more appropriate than linear model of the chosen (even the full path) features.

Table 1. Experimental result

Algorithm		AUC	
RWR [20]		0.806	
BiRW [22]		0.870	
RWRH [14]		0.886	
SRWR		Logistics	RBF
	FP	0.864	0.876
	FPH	0.886	0.885
	BP	0.868	0.872
	SAB	0.840	0.847

5   Conclusions

This paper focuses on the link prediction problem in social networks via the proposed Supervised Random Walk with Restart (SRWR). Though RWR has been intensively studied and applied to lots of real world domains, no research on its supervised version has been done before. Compared to traditional RWR algorithm, SRWR takes advantage of the exploitation of structural information encoded in the underlying problem, which makes it more adaptable and scalable than RWR. Inspired by the definition of RWR, the authors propose four path based features from the heterogeneous network named Generalized Bi-relational Network (GBN). GBN is a kind of heterogeneous network consists in two types of relations, namely **Similarity** and **Association**. Similarity relations only resides in the same type of nodes while association relations define the relations between different type of nodes. The nodes are classified into two category: query and candidate nodes. The link prediction problem is further defined to be the predication of links between query nodes and candidate nodes. GBN is appropriate for the modeling of most of link prediction tasks. The experiment is conducted on a chemical-disease-gene interaction network data set under the measurement of AUC gained, with 8 combinations of features and models. The result shows the superiority of SRWR to standard RWR. The best combination of SRWR achieves higher AUC value than BiRW - a state-of-the-art gene prioritization algorithm, and equal performance to RWRH - another cutting-edge gene prioritization algorithm in the literature.

Acknowledgement. This material is supported by National Institutes of Health under the grant number R01LM011986. The content of the information in this document does not necessarily reflect the position or the policy of the Government, and no official endorsement should be inferred. The U.S. Government is authorized to reproduce and distribute reprints for Government purposes notwithstanding any copyright notation here on. This work is also supported in part by the National High-Technology Research and Development Program (863 Program) of China under Grand 2013AA01A212, National Science Foundation Grant 61272067, 61370229 and Jiaying University Grant ("Collaboration Mechanism and Application in Social Networks.").

References

1. Backstrom, L., Leskovec, J.: Supervised random walks: predicting and recommending links in social networks. In: Proceedings of the Fourth ACM International Conference on Web Search and Data Mining, WSDM 2011, pp. 635–644. ACM, New York (2011)
2. Bromberg, Y.: Disease gene prioritization. PLoS Comput. Biol. **9**(4), e1002902 (2013). 00014
3. Chakrabarti, S., Agarwal, A.: Learning parameters in entity relationship graphs from ranking preferences. In: Fürnkranz, J., Scheffer, T., Spiliopoulou, M. (eds.) PKDD 2006. LNCS (LNAI), vol. 4213, pp. 91–102. Springer, Heidelberg (2006)
4. Chang, C.C., Lin, C.J.: LIBSVM: a library for support vector machines. ACM Trans. Intell. Syst. Tech. **2**(3), 27:1–27:27 (2011). 22106
5. Cohen, S., Kimelfeld, B., Koutrika, G.: A Survey on Proximity Measures for Social Networks. In: Ceri, S., Brambilla, M. (eds.) Search Computing. LNCS, vol. 7538, pp. 191–206. Springer, Heidelberg (2012)
6. Cukierski, W., Hamner, B., Yang, B.: Graph-based features for supervised link prediction. In: The 2011 International Joint Conference on Neural Networks (IJCNN), pp. 1237–1244, July 2011
7. Davis, A.P., Grondin, C.J., Lennon-Hopkin, K., Saraceni-Richards, C., Sciaky, D., King, B.L., Wiegers, T.C., Mattingly, C.J.: The comparative toxicogenomics database's 10th year anniversary: update 2015. Nucleic Acids Res. **43**(Database issue), D914–D920 (2015)
8. Fire, M., Tenenboim, L., Lesser, O., Puzis, R., Rokach, L., Elovici, Y.: Link prediction in social networks using computationally efficient topological features. In: 2011 IEEE Third International Conference on Privacy, Security, Risk and Trust (PASSAT) and 2011 IEEE Third Inernational Conference on Social Computing (SocialCom), pp. 73–80, October 2011
9. Hasan, M.A., Chaoji, V., Salem, S., Zaki, M.: Link prediction using supervised learning. In: Proceedings of SDM 2006 Workshop on Link Analysis. Counterterrorism and Security (2006). 00358
10. Hasan, M.A., Zaki, M.J.: A survey of link prediction in social networks. In: Aggarwal, C.C. (ed.) Social Network Data Analytics, pp. 243–275. Springer, USA (2011). 00107
11. Jaccard, P.: Étude comparative de la distribution florale dans une portion des Alpes et du Jura. Bulletin de la Societe Vaudoise des Sciences Naturelles **37**(142), 547–579 (1901)
12. Katz, L.: A new status index derived from sociometric analysis. Psychometrika **18**(1), 39–43 (1953)
13. Lao, N., Cohen, W.W.: Fast query execution for retrieval models based on path-constrained random walks. In: Proceedings of the 16th ACM SIGKDD International Conference on Knowledge Discovery and Data Mining, KDD 2010, pp. 881–888. ACM, New York (2010)

14. Li, Y., Patra, J.C.: Genome-wide inferring genecphenotype relationship by walking on the heterogeneous network. Bioinformatics **26**(9), 1219–1224 (2010)

15. Liben-Nowell, D., Kleinberg, J.: The link prediction problem for social networks. In: Proceedings of the Twelfth International Conference on Information and Knowledge Management, CIKM 2003, pp. 556–559. ACM, New York (2003)

16. Lu, L., Zhou, T.: Link prediction in complex networks: a survey. Physica A: Stat. Mech. Appl. **390**(6), 1150–1170 (2011). arXiv:1010.0725

17. Page, L., Brin, S., Motwani, R., Winograd, T.: The PageRank citation ranking: bringing order to the web (1999)

18. Pan, J.Y., Yang, H.J., Faloutsos, C., Duygulu, P.: Automatic multimedia cross-modal correlation discovery. In: Proceedings of the Tenth ACM SIGKDD International Conference on Knowledge Discovery and Data Mining, KDD 2004, pp. 653–658. ACM, New York (2004)

19. Salton, G.: Introduction to Modern Information Retrieval. Mcgraw-Hill College, New York (1983)

20. Tong, H., Faloutsos, C., Pan, J.Y.: Fast random walk with restart and its applications. In: Proceedings of the Sixth International Conference on Data Mining, ICDM 2006, pp. 613–622. IEEE Computer Society, Washington, DC, USA (2006)

21. Xia, J., Caragea, D., Hsu, W.: Bi-relational network analysis using a fast random walk with restart. In: Ninth IEEE International Conference on Data Mining, ICDM 2009, pp. 1052–1057 (2009). 00011

22. Xie, M., Hwang, T., Kuang, R.: Prioritizing disease genes by Bi-random walk. In: Tan, P.-N., Chawla, S., Ho, C.K., Bailey, J. (eds.) PAKDD 2012, Part II. LNCS, vol. 7302, pp. 292–303. Springer, Heidelberg (2012)

23. Zhang, J., Kong, X., Yu, P.S.: Predicting social links for new users across aligned heterogeneous social networks, October 2013. arXiv: arXiv:1310.3492 [physics]

FCL: A New Network Words Extraction Approach Based on Statistical Language Knowledge

Lili Mei, Heyan Huang, Xiaochi Wei, Peng Yuan, and Xian-Ling Mao[✉]

Beijing Engineering Research Center of High Volume Language Information
Processing and Cloud Computing Applications,
Department of Computer Science and Technology,
Beijing Institute of Technology, Beijing 100081, China
{lilymay,hhy63,wxchi,jackburd,maoxl}@bit.edu.cn

Abstract. New network words could benefit many NLP tasks such as
Chinese word segmentation and sentiment analysis. However, automatic
new network words extraction is a challenging task because new network
words usually have no fixed language pattern, and even appear with the
new meanings of existing words. To tackle these problems, this paper proposes a novel approach of FCL to extract new network words. It not only
considers domain specificity, but also combines with multiple statistical
language knowledge. First, we perform a filtering algorithm to obtain a
list of candidate new words. Then, we employ the statistical language
knowledge to extract the top ranked new network words. Experimental results show that our proposed approach is able to extract a large
number of new network words and notably outperforms the state-of-the-
art methods. Moreover, we also demonstrate our approach increases the
accuracy of word segmentation by 10 % on corpus containing new words.

Keywords: New network words extraction · Word segmentation ·
Domain specificity · Statistical language knowledge

1 Introduction

Unlike English and other western languages, Chinese texts have no space between
words. Therefore, word segmentation is a very important basic precursor in Chinese natural language processing (NLP). However, Chinese word segmentation
does not perform well on informal texts, e.g., Weibo or BBS, which is mainly
caused by widely distributed new words [1]. Among the new words, new network
words are the most common ones.

In Web 2.0 based social media, new network words are emerging every-
day. Homophonic words, typos and abbreviations are common phenomena in
user-generated content. Some widely used homophonic words and abbreviations
evolve into new meanings on social websites, and even the existing word may

© Springer Science+Business Media Singapore 2015
X. Zhang et al. (Eds.): SMP 2015, CCIS 568, pp. 119–130, 2015.
DOI: 10.1007/978-981-10-0080-5_11

Table 1. Examples of new network words

New Network Word	Origination	English Translation
女票	女朋友	girlfriend
涨姿势	长知识	knowledge have been increased
高富帅	个子高、富有、帅气	a tall, rich and handsome man
走召弓虽	超强	very strong

have different new explanation. For example, the Chinese new word "女票 (girl-friend)" is the abbreviation of "女朋友 (girlfriend)". Some examples of new network words are shown in Table 1.

New words extraction is indispensable to many NLP tasks such as Chinese word segmentation [2] and sentiment analysis [3]. However, automatic new words extraction is a challenging task. The reasons are as follows: (1) new words often have no fixed language pattern, appearing in a new form; (2) many new words appear with the new meanings and usages of existing words; (3) it is very difficult to identify low-frequency new words.

Existing methods for new words extraction have made significant progress. However, these methods are suffering from poor flexibility and portability [4–6], or they can not capture special features of new words [7–9]. To address these shortcomings, we consider domain specificity and combine with multiple statistical language knowledge to extract new words. Our main ideas are as follows: (1) It is intuitive that new network words rarely appear in the News corpus. Therefore, we extract n-grams from social website corpus and perform a filtering algorithm through the News corpus to obtain a list of candidate new words. (2) The statistical language knowledge can be used to quantify the possibility of a candidate new word being a new network word. So we introduce a ranking method for new network words extraction, considering word features like word frequency feature, word internal feature and neighborhood feature. Experimental results show that our method can effectively extract a large number of new network words. The main contributions of this paper are summarized as follows:

- We propose a novel approach on the task of new network words extraction. The approach is fully unsupervised which avoids the time-consuming labeling procedure. Furthermore, It requires no linguistic resources.
- We demonstrate the effectiveness of statistical language knowledge, such as string frequency, string cohesion and string liberalization, in the task of new network words extraction. No manually defined rule is needed to filter undesirable words.
- Experiments show that our proposed method increases the accuracy of word segmentation by 10 % on corpus containing new words.

The remainder of the paper is structured as follows: Sect. 2 summarizes the related work. In Sect. 3, we describe our method in detail. Then, the experimental results and discussions are presented in Sect. 4. Finally, we conclude our work and suggest future work.

2   Related Work

Extensive work has been done on new words extraction, which can be categorized as rule-based methods, statistical methods and hybrid methods. Our work falls into the statistical category.

In the rule-based methods, Isozaki generated and refined rules by decision tree learning [4]. By applying the refined rules, They got named entity candidates. Then non-overlapping candidates were selected by a kind of longest match method. Chen and Ma employed statistical and morphological rules to extract Chinese new words [5]. Meng et al. used parsing information to extract new words and the rules were built on the sentences [6]. For the rule-based methods, defining rules is difficult and the rules are often domain-specific which result in poor flexibility and portability.

In the statistical methods, Peng et al. considered word segmentation and new words extraction as a unified process [7]. They employed Conditional Random Fields (CRF) [10] to perform word segmentation. In [8], all potential unknown words were classified into single-character and affix model based on structures of unknown words. Then some filtration methods based on statistical information were performed. He and Zhu proposed a bootstrap method to extract new words [9]. Mutual information and Entropy were used. In [11,12], they treated new words detection as a binary classification problem. Xu et al. used the model of CRF [13]. In the Statistical methods, new words detection can be measured by Pointwise Mutual Information (PMI) [14], Independent Word Probability (IWP), Word Formation Power (WFP), Enhanced Mutual Information (EMI) [15] and multi-word expression distance (MED) [16]. The statistical methods suffer from the problem of low-frequency words and usually could not capture special features of new words well. Our work consider domain specificity and combine with multiple statistical language knowledge, which can overcome their disadvantages.

Hybrid methods are the combination of the rule-based methods and statistical methods. Huang et al. designed statistical measures to quantify the utility of lexical patterns and the extracted patterns could be further used in finding new words [3]. The shortcoming of this method is only extracting adjective new words. In [17], they proposed to use statistical information to provide the internal criteria, simultaneously employing rule-based methods to capture external criteria.

3   Our Approach

3.1   The Overview of Our Approach

For Chinese new network words extraction, our idea is inspired by the differences between social website corpus and News corpus. We collect user-generated contents from social website, e.g., Baidu Tieba, which is a good data source for extracting new network words. After the preprocessing, we firstly extract bi-grams, tri-grams, four-grams and five-grams from social website corpus. Then, a filtering algorithm based on domain specificity is performed to obtain a list of candidate new words. Finally, we introduce a ranking method for new network words extraction, based on statistical language knowledge like **string frequency**, **string cohesion** and **string liberalization**. In the following subsection, we present the above steps in detail.

3.2   A Filtering Algorithm Through the News Corpus

Firstly we extract bi-grams, tri-grams, four-grams and five-grams from social website corpus. After filtering the n-grams containing stop words, there are many normal words and garbage strings, such as "学校 (school)", "高跟鞋 (heels)" and "要考[1]", which are noise words for our new network words extraction. It is intuitive that new network words rarely appear in the News corpus, but the normal words and garbage strings are common phenomenon in the n-grams of News corpus. Therefore, we introduce a filtering algorithm through the News corpus, which is given in Algorithm 1.

Algorithm 1. A filtering algorithm through the News corpus

---

1 **Input**: The News corpus N, all the n-grams($n \in \{2,3,4,5\}$) extracted from social website corpus as set TG
2 **Output**: A list of candidate new words CW
3 Extract n-grams($n \in \{2,3,4,5\}$) from N as set NG
4 $CW = \Phi$
5 **foreach** n-gram named g in set TG **do**
6   **if** $g \in NG$ **then**
7         $\lfloor\ TG = TG - g$
8   **else**
9         $\lfloor\ CW = CW + g$
10 **return** CW

In Algorithm 1 firstly, the given News corpus is split into sentences according to the punctuation marks. Then, we extract n-grams from the News corpus as set NG. If our n-gram extracted from social website corpus occurs in NG, the gram will be neglected. Otherwise, it will be added into the candidate new words list. Finally, we get a list of candidate new words.

3.3   Measuring Word Features

After the filtering algorithm through the News corpus, there are many noise candidate new words, such as "富帅不 (see footnote 1)", "密达我 (see footnote 1)" and "大普奔 (see footnote 1)". Therefore, we should consider the special features of words. Our approach takes use of three kinds of statistical language knowledge to measure word features.

---

[1] It is a garbage string and has no meaning.

String Frequency. It is intuitive that a string can be potential word if it has high frequency. For example, the frequency of "高富帅 (a tall, rich and handsome man)" is generally high than the frequency of "富帅不 (see footnote 2)". When $S = s_1 s_2 \ldots s_n$ expresses a string, string frequency F is the number of S occuring in the corpus.

String Cohesion. String cohesion is the correlation of different components in $S = s_1 s_2 \ldots s_n$, which indicates word internal feature. If a string can be a potential word, it must have strong cohesion. For example, "思密达 (new modal particle)" is a better potential word than "密达我[2]" due to the strong cohesion. Enhanced mutual information [15] is a useful criterion to evaluate string cohesion, which is defined as the ratio of its probability of being a multi-character to its probability of not being a multi-character:

$$C(S) = \log_2 \frac{F(S)}{\prod\limits_{i=1}^{n}(F(s_i) - F(S))} \tag{1}$$

where $F(S)$ and $F(s_i)$ respectively denote the string frequency of S and s_i. The key idea of string cohesion is to measure a string's dependency of internal feature. The larger the value is, the more possible the expression will be a potential new word. Table 2 gives some examples of high string cohesion and low string cohesion.

Table 2. Examples of high string cohesion and low string cohesion

String	High Cohesion	String	Low Cohesion
楼主*	0.54	楼们**	0.07
思密达*	0.47	我个一**	0.09
萌妹纸*	0.42	我吧你**	0.08
喜大普奔*	0.39	为你比我**	0.14
深藏功与名*	0.41	我不就是想**	0.09

* new network word with high cohesion
** low-cohesion string with no meaning

String Liberalization. String liberalization indicates neighborhood feature of a string. If a string can be a potential word, it will be more commonly used with diversified neighborhood. That is to say, the word has high liberalization and can be used in many different linguistic scenarios. For example, "喜大普奔 (good news, and want to tell others)" is a better potential word than "大普奔 (see footnote 2)" due to the high string liberalization. This can be measured by information entropy, which is usually used to indicate the degree of uncertainty or randomness. If all the left characters of S are set $C_l = \{c_1, c_2, \ldots, c_l\}$ and all

---

[2] It is a garbage string and has no meaning.

the right characters of S are set $C_r=\{c_1, c_2, \ldots, c_r\}$, the left entropy and right entropy of S are as follows:

$$L_l(S) = -\sum_{i=1}^{l} \frac{F(c_i S)}{F(C_l S)} \times \log \frac{F(c_i S)}{F(C_l S)} \tag{2}$$

$$L_r(S) = -\sum_{i=1}^{r} \frac{F(S c_i)}{F(S C_r)} \times \log \frac{F(S c_i)}{F(S C_r)} \tag{3}$$

where $F(c_i S)$ and $F(S c_i)$ respectively denote the string frequency of $c_i S$ and $S c_i$. $F(C_l S)$ is the sum of $F(c_i S)$ ($c_i \in C_l$). It is the same with $F(S C_r)$. The key idea of string liberalization is to measure a string's diversity of neighborhood features. If the left or right neighbor of a string is contributed by a few fixed characters, the entropy will be low and the string liberalization is also very low. Table 3 gives some examples of left entropy and right entropy.

Table 3. Examples of left entropy and right entropy

String	High liberalization		String	Low liberalization	
	Left	Right		Left	Right
蛇精病*	0.28	0.65	精病**	0.03	0.62
脑残粉*	0.60	0.89	残粉**	0.002	0.86
韩国棒子*	0.47	0.70	韩国棒**	0.42	0.009
羡慕嫉妒恨*	0.38	0.46	慕嫉妒**	0	0.06
哭晕在厕所*	0.61	0.37	晕在厕**	0.06	0

* new network word with high liberalization
** low-liberalization string with no meaning

3.4   A Ranking Method for New Network Words Extraction

Considering the aforementioned features in Sect. 3.3, we combine string frequency, string cohesion and string liberalization together for new network words extraction. Thus, the possibility of new words can be formulated as follows:

$$FCL(w) = \alpha \hat{F}(w) + \beta \hat{C}(w) + \gamma \log \frac{\hat{L}_l(w) + \hat{L}_r(w)}{|\hat{L}_l(w) - \hat{L}_r(w)|^\sigma} \tag{4}$$

where $\alpha + \beta + \gamma = 1$. $\hat{F}(w)$ and $\hat{C}(w)$ are the normalized forms of string frequency and string cohesion respectively. $\hat{L}_l(w)$ and $\hat{L}_r(w)$ are the normalized forms of left entropy and right entropy. $\alpha, \beta, \gamma \in [0, 1]$ determines which type of features dominates new network words extraction. $\alpha = 0$ means the possibility of candidate new words is estimated by only considering string cohesion and string liberalization, ignoring the contribution of string frequency. Otherwise, when $\alpha = 1$, the possibility of candidate new words is estimated by only considering string frequency. β, γ have the similar roles with α. σ is used to adjust the impact of the difference by left entropy and right entropy. In our experiments, the best performance is obtained when $\alpha = 0.3, \beta = 0.4, \gamma = 0.3, \sigma = 0.1$.

After computing the FCL values of the candidate new words, we obtain a ranked list of all the words. In order to finish the new network words extraction, we set a threshold K. The top K new words will be considered to be the new network words. We denote our proposed method as FCL.

4   Experiments and Discussions

In this section, to evaluate our method, we conduct the following experiments: (1) we compare our method to several baselines; (2) we compare the impact of different features on new network words extraction; (3) we perform parameter tuning with extensive experiments; (4) we demonstrate how new network words benefit word segmentation.

4.1   Experiment Setup

Datasets: To evaluate our method, we used three datasets. (1) Baidu Tieba is a good social website data source for extracting new network words. Therefore, we crawled 3,524,584 **Tieba** posts[3] to evaluate our proposed method. These posts range from January to December of 2014. (2) The **News** corpus is provided by the NIST Open Machine Translation evaluation (OpenMT 2015) [18,19], which contains 9,517,292 sentences of xinhuanet[4]. (3) We crawled 1,0237,813 **Weibo** posts[5], applied in the word segmentation task. The Weibo posts were segmented into single words using a Chinese word segmentation tool named ICTCLAS [20].

The extracted new network words were manually annotated, where three annotators were involved. Two annotators were requested to judge whether the extracted word was a new network word. When conflicts occurred, the third annotator made final judgement. The annotation led to 1193 new network words.

Evaluation Metrics: We select precision (P), recall (R), f-measure (F) as metrics:

$$P = \frac{number\ of\ correct\ extraction}{total\ number\ of\ extraction}$$

$$R = \frac{number\ of\ correct\ extraction}{total\ number\ of\ new\ network\ words}$$

$$F = (2 * P * R)/(P + R)$$

---

[3] http://tieba.baidu.com/.
[4] http://www.xinhuanet.com/.
[5] http://weibo.com/.

4.2   Our Method vs. Different Baselines

To prove the effectiveness of the proposed **FCL** method, we select some methods for comparison as follows: New Word Detection(**NWD**) method [3], Mutual Information(**MI**) method used in [9] and Pointwise Mutual Information(**PMI**) method [14]. For the MI and PMI methods, we add string frequency and string liberalization into them and set the same parameters for the fairness of comparison. Figure 1 presents experimental results.

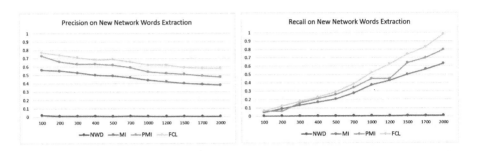

Fig. 1. Precisions and recalls of NWD, MI, PMI and FCL. X-axis is the top words threshold K, and Y-axis is precision or recall

Observing from Fig. 1, we can see that our method outperforms other methods both in precision and recall. It proves the effectiveness of the proposed method. For the NWD method, the performance is extremely dreadful, which is mainly because this method only extracts adjective words, ignoring other parts of speech. Besides, the segmentation tool can not perform well on the informal texts. Our method has much better performance than PMI and MI methods. We believe the reason is that EMI can measure a string's dependency of internal feature better than PMI and MI. We also notice that when K grows bigger, the precision decreases while the recall increases. That is because the bigger value of K can generate a wilder coverage, but bring in more noisy words. Clearly, when $K = 2000$, Our method could cover 98 % of the whole new network words set. Thus, it demonstrates our method is quite powerful in generating a large number of new network words.

4.3   Evaluation of Different Statistical Language Knowledge

In this subsection, we discuss which combination of word features is more effective for new network words extraction. For comparison, we design six baselines, noted as **F**, **C**, **L**, **F+C**, **F+L** and **C+L**. F only employs string frequency. C only employs string cohesion. L only employs string liberalization. F+C considers both string frequency and string cohesion. F+L and C+L have the similar way with F+C. Moreover, **F+C+L** is our method which considers all word features, referring to Eq. (4) with $\alpha = 0.3, \beta = 0.4, \gamma = 0.3$ and $\sigma = 0.1$. Table 4 presents experimental results when the top words threshold K varies.

Table 4. Results of different combinations of word features

Methods	K = 100			K = 300			K = 500			K = 1000			K = 1500			K = 2000		
	P	R	F	P	R	F	P	R	F	P	R	F	P	R	F	P	R	F
F	0.15	0.01	0.02	0.18	0.05	0.07	0.19	0.08	0.11	0.17	0.14	0.16	0.16	0.20	0.18	0.15	0.26	0.19
C	0.10	0.01	0.02	0.09	0.02	0.04	0.10	0.04	0.06	0.08	0.07	0.07	0.08	0.10	0.09	0.08	0.13	0.10
L	0.44	0.04	0.07	0.37	0.09	0.15	0.38	0.16	0.23	0.36	0.30	0.33	0.34	0.42	0.37	0.30	0.49	0.37
F+C	0.25	0.02	0.04	0.18	0.04	0.07	0.18	0.08	0.11	0.18	0.15	0.16	0.16	0.20	0.18	0.16	0.28	0.21
F+L	0.56	0.05	0.09	0.53	0.13	0.21	0.49	0.21	0.29	0.45	0.38	0.41	0.40	0.50	0.44	0.38	0.63	0.47
C+L	0.70	0.06	0.11	0.59	0.15	0.24	0.57	0.24	0.33	0.50	0.42	0.45	0.45	0.56	0.50	0.41	0.68	0.51
F+C+L	**0.77**	**0.06**	**0.12**	**0.71**	**0.18**	**0.28**	**0.69**	**0.29**	**0.41**	**0.62**	**0.52**	**0.57**	**0.59**	**0.74**	**0.66**	**0.59**	**0.98**	**0.73**

From results, we observe that F+C, F+L and C+L perform better than F and C. F+L and C+L perform better than L. These results indicate every word feature is necessary for new network words extraction. Moreover, F+C+L notably outperforms other baselines in all different K. It demonstrates combination of all different word features is effective.

4.4   Parameter Tuning

In this subsection, we discuss the variation of extraction performance when changing α, β, γ and σ in Eq. (4). Mean Reciprocal Rank (MRR) is an effective metric to test the extraction performance. Therefore, we select a list of annotated new words with size of n, named DIC. MRR is computed by $MRR = \frac{1}{n} \sum_{i=1}^{n} \frac{1}{p_i}$, where p_i is the ranking position of every new word in DIC. Experimental results are shown in Table 5 and Fig. 2.

Table 5 presents the MRR-values of new network extraction with varying α, β, γ and fixing $\sigma = 0.1$. Due to the space limitation, we only show twelve groups of parameter settings. we observe the best performance is obtained when $\alpha = 0.3, \beta = 0.4, \gamma = 0.3$. It indicates that string frequency, string cohesion and string liberalization are all useful for new network words extraction. The performance benefits from their combination. Figure 2 present the MRR-values with varying σ from 0 to 1 and fixing $\alpha = 0.3, \beta = 0.4, \gamma = 0.3$. We notice the performance increases when σ is set from 0 to 0.1. When σ gets bigger, performance, however, decreases. The best performance is achieved when $\sigma = 0.1$.

4.5   Application of New Network Words to Word Segmentation

In this subsection, we demonstrate whether new network words would benefit word segmentation. For this purpose, we randomly sampled 500 **Weibo** posts that contain at least one of our annotated new network words. We compare four different kinds of lexicons for word segmentation. One of the lexicons is the default lexicon (DL) in ICTCLAS [20]. Moreover, we add three resources into the default lexicon: the top 1000 words produced by our approach (denoted by T1000), all correct new words produced by our approach (denoted by CNW, including 1172 new words) and all annotated new words (denoted by ANW,

Table 5. The MRR-values with varying parameter α, β, γ

α	β	γ	MRR	α	β	γ	MRR
0.8	0.1	0.1	0.0050	0.1	0.1	0.8	0.0109
0.5	0.3	0.2	0.0119	0.2	0.3	0.5	0.0195
0.2	0.5	0.3	0.0198	0.3	0.5	0.2	0.0152
0.1	0.8	0.1	0.0026	0.3	0.3	0.4	0.0205
0.3	0.5	0.2	0.0152	**0.3**	**0.4**	**0.3**	**0.0218**
0.5	0.2	0.3	0.0178	0.4	0.3	0.3	0.0211

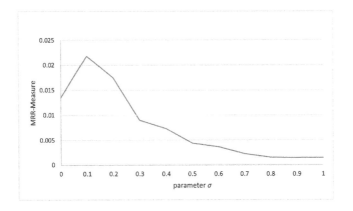

Fig. 2. The MRR-values with varying parameter σ

including 1193 new words), respectively. Thus, the four different kinds of lexicons are **DL, DL+T1000, DL+CNW** and **DL+ANW**. We use segmentation accuracy to evaluate the effect of four lexicons to word segmentation. Table 6 presents experimental results.

Table 6. The accuracy of word segmentation with four lexicons

Lexicon	Accuracy	Lexicon	Accuracy
DL	77.89 %	DL+CNW	89.17 %
DL+T1000	87.54 %	DL+ANW	89.60 %

We can see from Table 6 that all three lexicons which contain new words improve the performance remarkably. The lexicon DL+T1000 is generated by our method, which increases the performance of word segmentation by 10 %. The lexicon DL+CNW outperforms DL+T1000, which is mainly because T1000 may contain words that are not new words. We also observe that DL+ANW outperforms DL+CNW. The reason is that the number of annotated new words is bigger than the number of correct new words produced by our approach.

5 Conclusion and Future Work

In this paper, We propose a novel approach on the task of new network words extraction. The approach is fully unsupervised and purely data-driven. We perform a filtering algorithm based on domain specificity and employ the statistical language knowledge considering word frequency feature, word internal feature and neighborhood feature to extract new network words. No manually defined rule is needed to filter undesirable words. Compared to different baselines, experimental results prove the effectiveness of our approach. What's more, experiments also demonstrate new network words benefit word segmentation obviously.

The proposed method can not only extract new network words, but also extract domain new words like the domain words of computer science. In the future, we intend to combine our method with some known methods to extract domain new words. We are also considering how to excavate more useful features to improve the performance of new network words extraction.

Acknowledgment. This work was supported by the State Key Program of National Natural Science of China (Grant No. 61132009), the National High Technology Research and Development Program of China (863 Program, No. 2015AA015404) and the National Natural Science Foundation of China (No. 61201351 and 61402036).

References

1. Sproat, R., Emerson, T.: The first international Chinese word segmentation bake-off. In: Proceedings of the Second SIGHAN Workshop on Chinese Language Processing, vol. 17, pp. 133–143. Association for Computational Linguistics (2003)
2. Sun, X, Wang, H, Li, W.: Fast online training with frequency-adaptive learning rates for Chinese word segmentation and new word detection. In: Proceedings of the 50th Annual Meeting of the Association for Computational Linguistics, vol. 1, pp. 253–262. Association for Computational Linguistics (2012). Long Papers
3. Huang, M., Ye, B., Wang, Y., Chen, H., Cheng, J., Zhu, X.: New word detection for sentiment analysis. In: Proceedings of the 52nd Annual Meeting of the Association for Computational Linguistics, pp. 531–541, Baltimore (2014)
4. Isozaki, H.: Japanese named entity recognition based on a simple rule generator and decision tree learning. In: Proceedings of the 39th Annual Meeting on Association for Computational Linguistics, pp. 314–321 (2001)
5. Chen, K.J., Ma, W.Y.: Unknown word extraction for Chinese documents. In: Proceedings of the 19th International Conference on Computational Linguistics, vol. 1, pp. 1–7. Association for Computational Linguistics (2002)
6. Meng, Y., Yu, H., Nishino, F.: Chinese new word identification based on character parsing model. In: Proceedings of 1st IJCNLP, pp. 489–496, Hainan (2004)
7. Peng, F., Feng, F., McCallum, A.: Chinese segmentation and new word detection using conditional random fields. In: Proceedings of the 20th International Conference on Computational Linguistics. Association for Computational Linguistics, p. 562 (2004)

8. Jiang, X., Wang, L., Cao, Y., Lu, Z.: Automatic recognition of chinese unknown word for single-character and affix models. In: Wang, Y., Li, T. (eds.) ISKE2011. AISC, vol. 123, pp. 435–444. Springer, Heidelberg (2011)

9. He, S., Zhu, J.: Bootstrap method for Chinese new words extraction. In: 2001 IEEE International Conference on Acoustics, Speech, and Signal Processing, 2001. Proceedings, (ICASSP 2001), pp. 581–584. IEEE (2001)

10. Lafferty, J., McCallum, A., Pereira, F.C.N.: Conditional random fields: probabilistic models for segmenting and labeling sequence data. In: Proceedings of the 18th International Conference on Machine Learning, pp. 282–289 (2001)

11. Li, H., Huang, C.-N., Gao, J., Fan, X.: The use of SVM for chinese new word identification. In: Su, K.-Y., Tsujii, J., Lee, J.-H., Kwong, O.Y. (eds.) IJCNLP 2004. LNCS (LNAI), vol. 3248, pp. 723–732. Springer, Heidelberg (2005)

12. Guodong, Z.: A chunking strategy towards unknown word detection in chinese word segmentation. In: Dale, R., Wong, K.-F., Su, J., Kwong, O.Y. (eds.) IJCNLP 2005. LNCS (LNAI), vol. 3651, pp. 530–541. Springer, Heidelberg (2005)

13. Xu, Y., Wang, X., Tang, B., Wang, X.: Chinese unknown word recognition using improved conditional random fields. In: Eighth International Conference on Intelligent Systems Design and Applications, 2008, ISDA 2008, pp. 363–367. IEEE (2008)

14. Church, K.W., Hanks, P.: Word association norms, mutual information, and lexicography. In: Computational Linguistics, pp. 22–29 (1990)

15. Zhang, W., Yoshida, T., Tang, X., Ho, T.B.: Improving effectiveness of mutual information for substantival multiword expression extraction. In: Expert Systems with Applications, pp. 10919–10930 (2009)

16. Bu, F., Zhu, X., Li, M.: Measuring the non-compositionality of multiword expressions. In: Proceedings of the 23rd International Conference on Computational Linguistics, pp. 116–124. Association for Computational Linguistics (2010)

17. Wu, A., Jiang, Z.: Statistically-enhanced new word identification in a rule-based Chinese system. In: Proceedings of the Second Workshop on Chinese Language Processing, pp. 46–51. Association for Computational Linguistics (2000)

18. Liberman, M., et al.: Emotional Prosody Speech and Transcripts LDC2002S28. CD-ROM. Linguistic Data Consortium, Philadelphia (2002)

19. Huang, S., David, G., George, D.: Multiple-Tanslation Chinese Corpus LDC2002T01. Web Download File. Linguistic Data Consortium, Philadelphia (2002)

20. Zhang, H.P., Yu, H.K., Xiong, D.Y., Liu, Q.: HHMM-based Chinese lexical analyzer ICTCLAS. In: Proceedings of the Second SIGHAN Workshop on Chinese Language Processing, vol. 17, pp. 184–187. Association for Computational Linguistics (2003)

Systematic Comparison of Question Target Classification Taxonomies Towards Question Answering

Tianyong Hao[1,2], Wenxiu Xie[1], Chun Chen[1], and Yuming Shen[1(✉)]

[1] Cisco School of Informatics, Guangdong University of Foreign Studies, Guangzhou, China
{981555724, 1733697972}@qq.com, ymshen2002@163.com
[2] Key Lab of Language Engineering and Computing of Guangdong Province,
Guangdong University of Foreign Studies, Guangzhou, China
haoty@126.com

Abstract. Question target classification is one of the essential research topics in question answering. Accurate identification and classification of question targets can help understand questions for retrieving relevant passages and assist answer extraction and ranking for improving answer retrieval accuracy and user satisfaction with return answers. This paper presents a systematic analysis on question target classification taxonomy. We investigate existing definitions of the classification and propose a concise definition. We then compare the existing classification taxonomies. The relevancies of the taxonomies are analyzed, inspiring us to propose a new taxonomy classification strategy. We finally summarize the characteristics and tendency of the current research of question target classification taxonomy. The systematic comparison is expected to provide consistent and meaningful guidance in the research of question target understanding in question answering.

Keywords: Question target classification · QA · Taxonomy · Answer type

1 Introduction

The essential of a Question Answering (QA) system is the analysis and understanding of question, as well as answer extraction and ranking to obtain concise answers rather than a long list of documents [1, 2]. Therefore, correct identification of answer types is an important step since it is the key to understand users' question intention to extract needed answers. Targeting this purpose, Question Target Classification (QTC) is to represent the semantic classes of answers according to question target (also as answer type) so as to facilitate answer extraction [3].

Laokularat [3] summarized the significance of QTC in QA system as 3 points: (1) reduce the volume of candidate answers; (2) help review different question types and design corresponding solutions; (3) filter out irrelevant answers. Moreover, the correct identification of question target has been proved to be one of the most essential factors for the success of a QA system [4–7]. As addressed by Srihariet et al. [8], the answer type is important in QA process as the key to find correct answers to users' questions is correctly locating question target to identify the expected answer type. According to statistic by

© Springer Science+Business Media Singapore 2015
X. Zhang et al. (Eds.): SMP 2015, CCIS 568, pp. 131–143, 2015.
DOI: 10.1007/978-981-10-0080-5_12

Moldovan et al. [7], 36.4 % of QA failures are caused by incorrect question analysis, while 28.2 % of failures are related to answer type identification. Therefore, the research of QTC is a vital task in QA.

Though there is existing research on QTC, the task still has a number of difficulties for resolving: (1) QTC relies heavily on syntactic analysis, semantic analysis, and named entity recognition, etc. However, these methods are sensitive to languages and hard to be generalized [9]; (2) the contradiction of the number of classes and the classification accuracy. As addressed by Tran et al. [10], the increase of fine-grained classes can provide more semantic representations of question targets but increase the classification error rate as well; (3) word sense disambiguation problem existing in target classification; (4) the inconsistency of question target annotations caused by human understanding; (5) the incorrect extraction of target words caused by natural language processing methods or tools [12], e.g., syntactic tree generation.

There are comparisons among classification methods but lacks of deep analysis of classification taxonomy. This paper conducts a systematic investigation and comparison of the QTC taxonomies, aiming for providing evidences and suggestions for appropriate taxonomy design or application in QA system. We firstly investigate the definitions of QTC, clarifying the meaning of QTC. Afterwards, a new and concise definition is given. Through the analysis of existing QTC taxonomies, we summarize them into four kinds: (1) taxonomies based on question interrogative type, (2) taxonomies based on the question description style, (3) taxonomies based on the sematic interpretation of question target, and (4) taxonomies for restricted domain. We then compare the taxonomies and conclude founding characterizes. Particularly, we collect more than 30 QTC taxonomies and quantify them according to the category sharing. Finally, we develop a dynamic network to visualize the correlation among the taxonomies. All the data and the visualization are publicly available on our website to facilitate researchers on the research of QTC.

2   The Definition of Question Target Classification

There is no consistent definition of Question Target Classification (QTC), though the research of question answering exists for more than two decades. In order to clarify its definition, we investigated existing representative definitions, as shown in Table 1.

From the existing definitions of QTC, we summarize that the underline meanings are similar though the ways of problem defining are different. Through our analysis, the definitions share the following points: (1) The categories of QT are predefined as taxonomy; (2) The categories of QT are a list of answer types or semantic categories of answers; (3) the task is to classify questions into related categories. Based on these common points, particularly, the difference between QTC and question topical classification, we propose a concise definition of QTC as follows:

Definition: Question target classification is to assign questions into a list of predefined question target (answer type) categories.

Table 1. Representative definitions of question target classification

Research	Definitions
Li & Roth [13]	Question classification is a task that, given a question, maps it to one of the predefined k classes, which indicates a semantic constraint on the sought-after answer
Sundblad [4]	Question classification can loosely be defined as: given a question (represented by a set of features), assign the question to a single category or a set of categories (answer types)
Cai et al. [11]	Question classification (or question categorization) is studied for this purpose by classifying user queried questions into predefined target categories
Loni [2]	The task of a question classifier is to assign one or more class labels, depending. On classification strategy, to a given question written in natural language
Laokulrat [3]	The goal of question classification is to map a question into a category that represents the type of information that is expected to be present in the final answer
Niu [14]	Question classification is to find out a category that is most similar to a given question from all predefined categories
Wen [15]	Question classification is mainly to divide a given question into corresponding semantic category according to its answer type

At the same time, we review formalized definition expression of QTC. Sundblad [4] defined question classification as the task of assigning a boolean value to each pair $< q_j, c_i > \in Q \times C$, where Q is the domain of questions and $C = \{c_1, c_2,..., c_{|C|}\}$ is a set of predefined categories. Assigning $< q_j, c_i >$ to T or F indicates that q_j belongs or NOT belongs to the category c_i, respectively. The task is to make the unknown target function $\hat{\Phi}: Q \times C \rightarrow \{T, F\}$ approximate the ideal target function $\Phi: Q \times C \rightarrow \{T, F\}$, such that $\hat{\Phi}$ and Φ coincide as much as possible. Niu [14] defined that question classification is to find out a category that is most similar to a question among all predefined categories. It can be formulized through a mapping function: $G: X \rightarrow \{C_1, C_2,...,C_n\}$, where X is the domain of questions and $\{C_1, C_2,...,C_n\}$ refers to predefined categories. For a question $x \in X$, G maps x to a category C_i. From the definitions, the key is to establish a mapping relationship between questions and answers, to classify the questions into a pre-defined question target categories. Therefore, they have the same meaning with our definition above.

Question classification by target (as QTC) and by topic (question topical classification) are similar on classification application but are significant different in classification purpose. The former is used to classify questions according to the answer types, but the latter is to classify question in accordance with the question's related topics [11, 12, 16].

Although research on question topics may help question target identification, question target plays a more important role in determining the answer type. For example, as for two questions *"Where do most tourists visit in Cyprus?"* and *"What is worth seeing in Cyprus?"* [12], they have different topics: *"tourists, visit, Cyprus"* and *"worth seeing, Cyprus"*. However, their answer types are the same as "Location/Sites" according to QTC, namely tourism location of specific spots. The answer type is more helpful to filter out irrelative candidate answers, e.g., about the population of Cyprus.

3   Question Target Classification Taxonomy

QTC is a vital part of a QA system. A typical architecture of a QA system in the Text REtrieval Conference (TREC, http://trec.nist.gov/) can be found in [16]. In general, a QA system consists of three components [17]: question analysis, document retrieval, and answer extraction. We describe the first and the last components as they are relevant to the QTC task. The question analysis includes question preprocessing, question classification, and question extension. This component firstly parses a given question and analyzes its semantics then classifies the question into a matched answer type category. The answer extraction returns a list of candidate answers. Filtering irrelevant answers and keeping expected answers are an essential step to improve QA accuracy. The filtering is usually based on the identification of the question target category from a pre-defined answer type taxonomy therefore the QTC taxonomy has an important influence on answer extraction quality [9].

Due to lacking standards to classify QTC taxonomies, some researchers classified existing taxonomies into three kinds [1]: answer type-based taxonomies [13, 46, 47], semantic interpretation-based taxonomies [44, 48, 49], and hybrid information-based taxonomies [47, 50, 51]. However, this kind of classification strategy has difficulty in distinguishing hybrid information and the other two kinds. Moreover, it is incapable of covering taxonomies that are defined by question manners and domains.

We have investigated nearly 40 QTC taxonomies from 1999 to 2015. Based on contrastive analysis (will be described in Sect. 4) on these taxonomies, we empirically propose a taxonomy classification strategy as follows: (1) taxonomies based on question interrogative type, (2) taxonomies based on the question description style, (3) taxonomies based on the sematic interpretation of question target, and (4) taxonomies for restricted domain. The four kinds cover the existing taxonomies, highlighting distinct features of the taxonomies and providing a clear distinguish guideline among them to facilitate the contrastive analysis of QTC taxonomies.

(1) Taxonomies based on interrogative type. This kind of taxonomies is based on common question interrogative type. For example, Oard et al. [52] proposed a list of interrogative types as "5 W + 3H" and Mudgal et al. [20] proposed a "6 W + 1H", which contains 7 coarse-grained categories including "Who", "Where", "What", "When", "Which", "Why", and "How" as well as 27 fine-grained categories including "Person", "Organization", etc.

(2) Taxonomies based on question description style. This kind of taxonomies is based on the manner or style of question asking. For example, Graesser &

Person [18, 19] proposed a QTC taxonomy including 18 categories such as "Verification", "Disjunctive", "Quantification", etc. Usually, it only contains single layer categories (coarse-grained categories).

(3) Taxonomies based on the sematic interpretation of question target. This kind of taxonomies is based on the sematic interpretation of question target (or expected answer type). The semantic interpretation can be in multiple layers, e.g., both Zhang & Lee [21] and Li & Roth [13] designed taxonomy in two layers.

(4) Taxonomies for restricted domain. This kind of taxonomies is based on a specific domain. For example, Su et al. [22] proposed a taxonomy for the tourism domain including 4 coarse-grained categories (e.g., "Scenic spot") and 18 fine-grained categories (e.g., "place position").

According to statistics from the QA Track of TREC, the kind (3) is being widely used one among the four kinds of taxonomies. This is probably due to the relative rich semantic interpretation of answer types and relative formal relation for semantic-based computing. We only list a taxonomy by Li & Roth [13] in Table 2 as example for kind (3) due to space limitation.

Table 2. The QTC taxonomy by Li & Roth [13] as example in our taxonomy classification

Coarse-grained	Fine-grained
Abbrev	Abbreviation, Expression abbreviated
Entity	Animal, Body, Color, Creative, Currency, Disease and medicine, Event, Food, Instrument, Language, Letter, Other, Plant, Product, Religion, Sport, Substance, Symbol, Technique, Term, Vehicle, Word
Description	Definition, Description, Manner, Reason
Human	Group, Individual, Title, Description
Location	City, Country, Mountain, Other, State
Numeric	Code, Count, Date, Distance, Money, Order, Other, Period, Speed, Temperature, Size, Weight

4 Comparison and Analysis

The classification of the existing taxonomies into four kinds enables us to horizontally compare taxonomy differences as well as compare accuracy differences. Through our deep analysis of the 38 QTC taxonomies in the last 16 years, we found a list of characteristics and summarized them as follows:

(1) **Language embodied.** We identified that QTC taxonomies vary from languages. From the comparison between English and Chinese classification taxonomies, all the investigated Chinese taxonomies removed the English coarse-grained category "Abbreviation". This is due to the fact that the abbreviations in English are usually

presented as short strings in capital letters, while there is no such linguistic phenomenon in Chinese language. Also, most Chinese classification taxonomies add "Time" category as an independent coarse-grained category. This is because that the expressions of time in English are basically fixed numbers or words, e.g., Thursday. Therefore, the "Time" category is organized as fine-grained category within "Numeric". However, in Chinese, the expressions of time are not limited to year, month, hour or minute, and the like. They can be more complex Chinese characters (such as "壹拾贰"). Due to the language characteristics, regardless of the popularity of some English classification taxonomies, e.g., developed by Li et al. [13], the revision of classification taxonomy according to language characteristics is needed on the application to other languages.

(2) **Open domain vs restricted domain.** Some scholars have focused on the customization of classification taxonomy in restricted domains, establishing appropriate classification taxonomies to improve the accuracy of QTC through the analysis of domain characteristics. However, according to QTC performance comparison of the existing taxonomies, taxonomies for restricted domain have not demonstrated obvious accuracy advantage. Furthermore, due to the category modification for restricted domain, some problems such as poorer universality and narrower adaptability of the taxonomies may occur. Particularly, additional domain identification step is of need in question analysis so as to confirm whether a question belongs to the domain, otherwise the accuracy of answer extraction may be potentially affected and decreased.

By comparing the open domain and restricted domain, we found that some commonly used categories are also included in restricted domain, e.g., "number", "time", and "location". However, with more and more researchers focusing on restricted domain research, the using of taxonomy tends to be specialization. For example, in the medical domain, Robert et al. [45] proposed a new classification taxonomy in 2015, with 14 main categories defined, i.e., "Anatomy", "Cause", "Complication", "Diagnosis", "Information", "Management", "Manifestation", "Other Effect", "Person Org", "Prognosis", "Susceptibility", "Other", "Not Disease", and "Research", targeting toward disease questions. Such study demonstrates the refinement development trend of classification taxonomy in restricted domain.

(3) **Semantic hierarchy of answer vs concept hierarchy of ontology.** Classification taxonomy based on the semantic hierarchy of answer is manifested as the hierarchy of coarse-grained and fine-grained categories with semantic relations. For example, there is "IS_A" relation, e.g., hyponymy between "Animal" and "Entity". From the perspective, the semantic hierarchy is similar to the concept relationship representations within ontology, top-level ontology in particular, e.g., The Suggested Upper Merged Ontology (SUMO)[1]. Some studies also regard the classification hierarchy of answer directly as a domain knowledge classification taxonomy, e.g., Li et al. [23] stated that classification categorizes the questions with same focus into domain concept of objects taking question mark as focus in domain ontology. Therefore,they

---

[1] http://www.adampease.org/OP/.

believed that question classification is a process of mapping question focuses to a domain ontology. However, the semantic hierarchy of answer is not truly equal to concept relationship of ontology. For example, the commonly used categories "distance" and "size" have the hyponymy "Attribute" in WordNet[2] rather than "Numeric" in QTC taxonomies. Besides, some words in questions are represented in the form of numeric, e.g., "1.5 km", and there exists semantic relationship, but not direct concept hyponymy. If applying concept hierarchy of ontology like WordNet, diversity in terms of different hierarchies will be occurred, thus making it difficult to correctly identify the type of answer. Therefore, QTC should also take into account both the expression characteristic of answers and semantic hierarchy of answers.

(4) **The performance comparison of classification taxonomies.** From the application perspective of classification taxonomies in Chinese and English language, the taxonomies based on the semantic hierarchy of answer types are more widely used in published academic articles. Particularly, most of QA answer systems partici- pated in TREC adopt this type of classification, demonstrating its wide acceptance in mainstream research. We therefore focus on the systematic investigation of the semantic hierarchy of QTC with 27 classification taxonomies investigated. Among them, 55.6 % of taxonomies apply single-layer semantic hierarchy (coarse-grained categories only), and 44.4 % apply double-layers hierarchy (coarse-grained and fine-grained categories). We sort out the evaluation results of QA systems based on the QTC taxonomies in TREC by referencing to relevant literature [24]. The performance comparison in terms of MRR is summarized as Table 3.

During the comparison, we find that double-layer taxonomies increase the difficulty of QTC classification and thus reduce classification accuracy slightly. Furthermore, the number of categories in the same layer also affects the accuracy. Generally speaking, the more the categories in need of mapping, the lower the classification accuracy obtained. In contrast, more semantic layers and categories can enhance the explicitness of QT, potentially providing better answer filtration to achieve higher QA accuracy. However, according to the evaluation results from TREC, the accuracy of some QA systems based on double-layer taxonomies exceed single- layer based systems because of different classification methods, e.g., Cymfony [41] system based on 18(22+) categories achieved best MMR as 0.66 in TREC 8.

Considering the better semantic representation of double-layer taxonomies in the advantage of answer extraction, we rank the classification taxonomies with double- layer hierarchy according to MMR performance. The following four QTC taxon- omies achieve highest scores: Cymfony (18/22+), ISI (7/140+), SMU (16/30+) and UIUC (6/50). Particularly, there are more coarse-grained categories in Cymfony and SMU, and the fine-grained categories in ISI are extremely rich. UIUC taxonomy contains more formally-defined semantic relations as hierarchy so that it can be mapped to ontology more conveniently, enhancing its extensibility. Furthermore, there is rare study on three-layer hierarchy due to the computation complex and accuracy decrease to some extent.

---

[2] http://wordnet.princeton.edu/.

Table 3. The performance comparsion of QA systems using different QTC taxonomies

QA systems	Coarse-grained	Fine-grained	MRR	
			50-byte limit	250-byte limit
UAlicante [32]	2	9	0.23(TREC9)	0.36(TREC9)
AT&T [27]	8	N/A	0.36(TREC8)	0.55(TREC8)
CL Research [25]	6	N/A	N/A	0.28(TREC8)
Cymfony [41]	18	22+	0.66(TREC8)	N/A
UFudan [30]	11	N/A	0.20(TREC9)	0.34(TREC9)
IBM(Ittycheriah) [37]	6	31	0.29(TREC9)	0.46(TREC9)
ISI [39]	7	140+	0.32(TREC8)	N/A
KAIST [26]	6	N/A	0.21(TREC9)	0.33(TREC9)
Korea [42]	25	46	0.37(unknown)	N/A
QALC [35]	5	17	0.18(TREC9)	0.41(TREC9)
UIowa [25]	4	N/A	0.02(TREC8)	0.06(TREC8)
Qanda [34]	5	10+	0.281(TREC8)	0.43(TREC8)
LASSO [53]	15	N/A	0.56(TREC8)	0.65(TREC8)
			0.58(TREC9)	0.76(TREC9)
UMontreal [31]	13	N/A	0.18(TREC8)	0.37(TREC9)
UWaterloo [33]	5	8+	0.32(TREC9)	0.46(TREC9)
NMSU [44]	27	N/A	0.22(TREC8)	0.27(TREC8)
NTT [36]	5	28	0.26(TREC8)	0.37(TREC8)
			0.23(TREC9)	0.39(TREC9)
NUS [38]	7	54	0.22(TREC8)	N/A
			0.27(TREC9)	
UMaryland [51]	6	N/A	0.30(TREC8)	N/A
QuASM [52]	7	N/A	0.19(TREC8)	0.34(TREC8)
				0.34(TREC9)
NSIR [55]	20	N/A	0.2(TREC8)	N/A
IBM(Prager) [54]	17	N/A	0.32(TREC9)	0.42(TREC9)
USheffield [28]	8+	N/A	0.08(TREC8)	0.11(TREC8)
QUALIFIER [40]	16	30+	0.8(TREC11)	N/A
Sun Microsystems [43]	7	N/A	N/A	0.34(TREC9)
UIUC [21]	6	50	0.18(unknown)	N/A
Xerox [29]	10	N/A	0.32(TREC8)	0.45(TREC8)
			0.23(TREC9)	0.35(TREC9)

(5) **The correlation between classification taxonomies.** We investigate 28 taxonomies with semantic hierarchy for open domain QA systems in total, including UIowa, AT&T, IBM(Ittycheriah), ISI, UIUC, and NUS. For each taxonomy, we respectively list out its coarse and fine-grained categories and then preprocess the categories by similar category combination, e.g., the categories "Numeric" and "number" are merged as "Numeric/Number". Afterwards, the categories are registered for analyzing the usage of every category among taxonomies. Figure 1 shows the sharing proportion of categories across all taxonomies. The most five sharable categories are "Location" (77 %), "Numeric/Number" (63 %), "Time" (63 %), "Person" (60 %), and "Organization" (57 %). All the taxonomies and the statistical result can be downloaded from our website[3].

In addition, to analyze the correlation between different QTC taxonomies, we apply Social Network Analysis [56] technique to automatically generate dynamic connection of the QTC taxonomies. As shown in Fig. 2, the larger (orange) points stand for the taxonomies while the smaller (blue) points represent specific categories. From the result, the larger points (taxonomies) sharing more smaller points (categories) tend to gather together. The larger points close to network center present that the taxonomies share more categories with other taxonomies, while the smaller points close to center denote that the categories are shared by more taxonomies, indicating the better universality. The dynamic network is available on our website[4], where each point can be dragged by mouse to view all the categories associated with current taxonomy or to view all the taxonomies sharing current category.

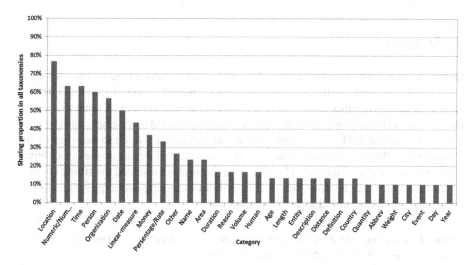

Fig. 1. The sharing proportion of categories across all QTC taxonomies

---

[3] http://www.zhukun.org/haoty/resources.asp?id=QTC_taxonomies.
[4] http://www.zhukun.org/haoty/resources.asp?id=QTC_taxonomies_network.

Fig. 2. The visualization of QTC taxonomies utilizing social network analysis technique (Color figure online)

5   Summary

Question target classification is able to facilitate question understanding thus is helpful for improving answer extraction quality, thus is one of essential parts of question answering research. Since there is no consistent definition and lack of deep analysis of question target classification taxonomy, this paper conducted a systematic analysis and comparison of the QTC taxonomies. We creatively found five characteristics of the differences of these taxonomies. Particularly, we compared the accuracies of different taxonomies on standard dataset. The sharing analysis also presented the strong connections among those taxonomies, which is helpful for designing and implementing QTC taxonomies for a new question answering system.

Acknowledgements. The work was supported by National Natural Science Foundation of China (grant No. 61403088, No. 61305094 and No. 61103169).

References

1. Feng, X.B.: Research on Question Classification and Keyword Extension of Chinese QA System (in Chinese). Doctorate Dissertation (2011)

2. Loni, B.: A survey of state-of-the-art methods on question classification. Delft University of Technology, Technical report (2011)
3. Laokulrat, N.: A survey on question classification techniques for question answering. KMITL Inf. Technol. J. **2**(1) (2013)
4. Sundblad, H.: Question classification in question answering systems. Linköping Institute of Technology at Linköping University, Thesis (2007)
5. Ittycheriah, A., Franz, M., Zhu, W.J., et al.: IBM's Statistical Question Answering System. In: TREC, pp. 229–235 (2000)
6. Hovy, E., Gerber, L., Hermjakob, U., et al.: Toward semantics-based answer pinpointing. In: International Conference on Human Language Technology Research, pp. 1–7 (2001)
7. Moldovan, D., Paşca, M., Harabagiu, S., et al.: Performance issues and error analysis in an open-domain question answering system. ACM Trans. Inf. Syst. (TOIS) **21**(2), 133–154 (2003)
8. Srihari, R., Niu, C., Li, W.: A hybrid approach for named entity and sub-type tagging. In: Conference on Applied Natural Language Processing, pp. 247–254 (2000)
9. Solorio, T., Pérez-Coutino, M., Montes-y-Gémez, M., et al.: A language independent method for question classification. In: COLING (2004)
10. Tran, D.H., Chu, C.X., Pham, S.B., et al.: Learning based approaches for vietnamese question classification using keywords extraction from the web. In: IJCNLP, pp. 740–746 (2013)
11. Cai, L., Zhou, G., Liu, K., et al.: Large-scale question classification in cQA by leveraging wikipedia semantic knowledge. In: CIKM, pp. 1321–1330 (2011)
12. Huang, Z., Thint, M., Qin, Z.: Question Classification using Head Words and Their Hypernyms. In: EMNLP, pp. 927–936 (2008)
13. Li, X., Roth, D.: Learning question classifiers. In: COLING, vol. 1, pp. 1–7 (2002)
14. Niu, Y.Q.: Study on Question Classification in Chinese Question Answering System (in Chinese). Master Thesis (2011)
15. Wen X.: Research on Question Classification and Candidate Answer Sentences Extraction in Chinese Question Answering System (in Chinese). Master Thesis (2006)
16. Gao, Y.Y.: Research on the Question Classification in Chinese Question Answering System (in Chinese). Master Thesis (2011)
17. Zheng, S.F., Liu, T., Qin, B., et al.: Overview of question-answering (in chinese). J. Chin. Inf. Process. **6**(6), 44–52 (2002)
18. Graesser, A.C., Louwerse, M.M., Burger, J., et al.: Question Generation and Answering Systems: R&D for Technology-enabled Learning Systems. Research roadmap for Federation of American Sciences, pp. 1–74 (2003)
19. Graesser, A.C., Person, N.K.: Question Asking During Tutoring. Am. Educ. Res. J. **31**(1), 104–137 (1994)
20. Mudgal, R., Madaan, R., Sharma, A.K., et al.: A Novel Architecture for Question Classification based Indexing Scheme for Efficient Question Answering. arXiv preprint arXiv:1307.6937 (2013)
21. Zhang, D., Lee, W.S.: Question classification using support vector machines. In: SIGIR, pp. 26–32 (2003)
22. Su, L., Yu, Z., Guo, J., et al.: Domain adaptation for question classification. J. Comput. Inf. Syst. **7**(9), 3261–3267 (2011)
23. Li, X.: Question Classification in Question Answering System (in Chinese). Doctorate Dissertation (2007)
24. Eichmann, D., Srinivasan, P.: Filters, webs and answers: the university of Iowa TREC-8 Results. In: TREC, pp. 259–266 (1999)

25. Litkowski, K.C.: Question-answering using semantic relation triples. In: TREC, pp. 349–356 (1999)

26. Lee, K.S., Oh, J.H., Huang, J.X., et al.: TREC-9 Experiments at KAIST: QA, CLIR and Batch Filtering. In: TREC, pp. 317–330 (2000)

27. Singhal, A., Abney, S.P., Bacchiani, M., et al.: AT&T at TREC-8. In: TREC, pp. 317–330 (1999)

28. Scott, S., Gaizauskas, R.J.: University of Sheffield TREC-9 Q&A System. In: TREC, pp. 707–716 (2000)

29. Hull, D.A.: Xerox TREC-8 Question Answering Track Report. In: TREC, pp. 743–753 (1999)

30. Wu, L., Huang, X., Zhou, Y., et al.: FDUQA on TREC 2003 QA task. In: TREC, pp. 246–253 (2003)

31. Laszlo, M., Kosseim, L., Lapalme, G.: Goal-driven answer extraction. In: TREC, pp. 563–573 (2000)

32. Vicedo, J.L., Llopis, F., Ferrández, A.: University of alicante experiments at TREC 2002. In: NIST Special Publication SP, vol. 251, pp. 595–602 (2003)

33. Clarke, C.L.A., Cormack, G.V., Kisman, D.I.E., et al.: Question answering by passage selection (multitext experiments for TREC-9). In: TREC, pp. 673–684 (2000)

34. Breck, E., Burger, J.D., Ferro, L., et al.: A sys called qanda. In: TREC, pp. 499–507 (1999)

35. Ferret, O., Grau, B., Illouz, G., et al.: QALC-the question-answering program of the language and cognition group at LIMSI-CNRS. In: TREC, pp. 465–475 (1999)

36. Takaki, T.: NTT DATA TREC-9 question answering track report. In: TREC, pp. 399–407 (2000)

37. Ittycheriah, A., Franz, M., Roukos, S.: IBM's statistical question answering system-TREC-10. In: TREC, pp. 258–265 (2001)

38. Zhang, D., Lee, W.S.: Web based pattern mining and matching approach to question answering. In: TREC, vol. 2, p. 497 (2002)

39. Hovy, E., Hermjakob, U., Ravichandran, D.: A question/answer typology with surface text patterns. In: International Conference on Human Language Technology Research, pp. 247–251 (2002)

40. Yang, H., Chua, T.S., Wang, S., et al.: Structured use of external knowledge for event-based open domain question answering. In: SIGIR, pp. 33–40 (2003)

41. Srihari, R., Li, W.: Information Extraction Supported Question Answering. Cymfony Net Inc., Williamsville (1999)

42. Han, K.S., Chung, H.J., Kim, S.B., et al.: Korea university question answering system at TREC 2004. In: TREC (2004)

43. Woods, W.A., Green, S., Martin, P., et al.: Halfway to question answering. In: TREC, pp. 489–501 (2000)

44. Ogden, B., Cowie, J., Ludovik, E., et al.: CRL's TREC-8 systems cross-lingual IR, and Q&A. In: TREC, pp. 513–522 (1999)

45. Roberts, K., Masterton, K., Fiszman, M., et al.: Annotating question types for consumer health questions. In: LREC Workshop on Building and Evaluating Resources for Health and Biomedical Text Processing (2014)

46. Zhang, Y., Liu, T., Wen, X.: Modified bayesian model based question classification. J. Chin. Inf. Process. **19**(2), 100–105 (2005). (in Chinese)

47. Lu, Z.J., Zhang, D.M.: Case-based question analyzer. J. Comput. Simul. **21**(5), 162–164 (2005). (in Chinese)

48. Soubbotin, M.M.: Patterns of potential answer expressions as clues to the right answers. In: TREC, pp. 293–303 (2001)

49. Du, Y., Huang, X., Li, X., et al.: A novel pattern learning method for open domain question answering. In: Natural Language Processing–IJCNLP, pp. 81–89 (2005)
50. Hovy, E., Hermjakob, U., Ravichandran, D.: A question/answer typology with surface text patterns. In: International Conference on Human Language Technology Research, pp. 247–251 (2002)
51. Kwok, C., Etzioni, O., Weld, D.S.: Scaling question answering to the web. ACM Trans. Inf. Syst. (TOIS) **19**(3), 242–262 (2001)
52. Oard, D.W., Wang, J., Lin, D., et al.: TREC-8 Experiments at Maryland: CLIR, QA and Routing. Maryland University College Park Library School (2000)
53. Pinto, D., Branstein, M., Coleman, R., et al.: QuASM: a system for question answering using semi-structured data. In: ACM/IEEE-CS Joint Conference on Digital Libraries, pp. 46–55 (2002)
54. Moldovan, D., Harabagiu, S., Pasca, M., et al.: LASSO: a tool for surfing the answer net. In: TREC, pp. 175–184 (1999)
55. Radev, D., Fan, W., Qi, H., et al.: Probabilistic question answering on the web. J. Am. Soc. Inf. Sci. Technol. **56**(6), 571–583 (2005)
56. Wölfer, R., Faber, N.S., Hewstone, M.: Social network analysis in the science of groups: cross-sectional and longitudinal applications for studying intra-and intergroup behavior. Group Dyn.: Theory, Res. Pract. **19**(1), 45–61 (2015)

Combining Convolutional Neural Network and Support Vector Machine for Sentiment Classification

Yuhui Cao, Ruifeng Xu$^{(\boxtimes)}$, and Tao Chen

Shenzhen Engineering Laboratory of Performance Robots at Digital Stage,
Harbin Institute of Technology Shenzhen Graduate School, Shenzhen, China
{caoyuhuiszu,xuruifeng.hitsz,chentao1999}@gmail.com

Abstract. In recent years, the classifiers based on convolutional neural network (CNN) and word embedding achieved good performances in sentiment classification tasks. However, the CNN-based model simply uses a fully connected layer for classification and it cannot perform a non-linear classification efficiently compared to the support vector machine (SVM) classifier. Target to this problem, in this paper, we combine CNN and SVM for sentiment classification. Firstly, continuous bag of word (CBOW) model is applied to construct word embedding. CNN is then utilized to learn feature vector representation corresponding to each sentence. The learned vector representations are fed to a SVM classifier as features for sentiment classification. Evaluations on the NLPCC2014 Sentiment Classification with Deep Learning Technology Task datasets (in short, NLPCC-SCDL) show that our model outperforms the top system in the NLPCC 2014 evaluation, on both English and Chinese sides.

Keywords: Sentiment analysis · Convolutional neural network · Support vector machine

1   Introduction

With the rapid development of E-commerce, more and more people purchase items and share their reviews online. These increasing reviews are valuable to new customer for making purchase decision and to the manufactory for improving their products. Thus, the sentiment analysis techniques, which identifies the subjective comments from product review texts and determines their polarities, attracts much research interests in recent years. The majority of sentiment analysis may be divided into subjectivity classification [14] and sentiment polarity classification [16,23]. More specifically, sentiment polarity classification contains binary classification (positive and negative) [16] and multivariate classification [15,23].

Generally speaking, existing works on sentiment classification may be camped into two major approaches, namely sentiment knowledge based approach [17] and machine learning based approach [16]. The former approach mainly utilizes the sentiment lexicon, rules and pattern matching for sentiment analysis [19,20].

© Springer Science+Business Media Singapore 2015
X. Zhang et al. (Eds.): SMP 2015, CCIS 568, pp. 144–155, 2015.
DOI: 10.1007/978-981-10-0080-5_13

Attributes to the increasing popularity of network informal language and flexible use of regular words in online product reviews, this approach has shown their limitations. The machine learning based approaches attracted increasing attentions in recent years. This approach normally uses words, Bi-grams, sentiment words, part-of-speech and information gain as features to represent the training samples. The machine learning based algorithms, such as support vector machines (SVMs), Naive Bayes (NB) and Maximum Entropy (ME), are applied to the feature vectors corresponding to the labeled dataset for training the classifier. Thus, feature engineering, a method learning features from texts, plays an important role. In recent years, the classifier based on convolutional neural network (CNN) and word embedding achieved good performances in sentiment classification. Unlike the bag of words representation, word embedding is distributed representation based on a neural probabilistic language model, which is expected to alleviate data sparseness because they are low dimensional, dense and continuous. However, the CNN-based model simply uses a fully connected layer for classification and it cannot perform a non-linear classification efficiently.

In this study, we present a CNN-SVM combined model for sentiment classification. In this model, continuous bog of word (CBOW) model is employed to construct word embedding. The CNN model is then applied to learn feature vector representations for the labeled training data. The learned feature vectors are fed to train the SVM classifier. Such a combined model is expected to combine the advantages of CNN model on feature learning and SVM model on efficient non-linear classification. Evaluations on NLPCC2014 Sentiment Classification with Deep Learning Technology Task datasets (NLPCC-SCDL) which is a sentiment labeled product review dataset, show that our combined model outperforms the CNN model and the top submitted system in the NLPCC 2014 evaluation, on both Chinese and English side. Our model achieves the highest known performance on this dataset, based on our knowledge, which shows the effectiveness of our CNN-SVM combined sentiment classification model.

The rest of this paper is organized as follows. Section 2 briefly reviews the related work. Section 3 presents the design and implementation of our CNN-SVM combined model for sentiment analysis. Section 4 gives the evaluation results and discussions. Finally, Sect. 5 gives the conclusions.

2 Related Work

2.1 Sentiment Analysis

Sentiment knowledge based approach uses universal [10] and domain specific [21] sentiment word lexicon, or sentiment rules and patterns to discriminate sentiment polarity. Kim et al. [10] used WordNet and HowNet as sentiment knowledge to classify sentiment polarity. Tong [21] constructed a domain specific sentiment lexicon for movie reviews by manually choosing sentiment phrases. Besides, D. Tang et al. [19] proposed to build a large-scale sentiment lexicon from Twitter by following a representation learning approach and cast sentiment lexicon learning as a phrase-level sentiment classification. Strfano Baccianella et al. [1]

constructed SentiWordNet by enriching WordNet with sentiment information for sentiment classification. As for Chinese sentiment lexicons, Minlie Huang et al. [6] proposed a method for detecting new sentiment words by exploring the frequent sentiment word patterns.

Machine learning based approach employs the machine learning models to learn features of texts and use discriminate methods to classify sentiment polarity. Pang et al. [16] proposed to classify movie reviews into positive/negative by using three different classifier including Naive Bayes, Maximum Entropy and SVM. They tested different feature combinations including unigrams, unigrams+bigrams and unigrams+POS (part-of-speech) tags. SVM can be used for both binary and multiple category classification [7]. Yang and Liu [24] compared SVM with linear Least-squares, Neural Network, Naive Bayes and k-nearest neighbors for sentiment classification. They found that SVM achieved an equal performance like other classifiers in their experiments.

2.2 Deep Neural Networks

Deep neural networks models have achieved remarkable results in computer vision [12] and speech recognition [5] area. In natural language processing area, many researchers used deep neural networks to learn word embedding [13] and perform composition over the learned word embeddings for classification [4]. Bengio et al. proposed a feed-forward neural network with a linear projection layer and a non-linear hidden layer to construct a neural language model [2]. The Collobert and Weston (C&W) model [3] is another neural language model based on the syntactic context of words. It substitutes the center word of a sentence by a random word to generate a corrupted sentence as a negative sample. Mikolov et al. [13] proposed a neural language model to learn word distribution representations including words semantics.

CNN model, which is initially invented for computer vision, has shown effective to natural language processing applications such as semantic parsing [25], search query retrieval [18], sentence modeling [8] and other traditional natural language processing tasks [4]. Kalchbrenner et al. [9] proposed a dynamic convolutional neural network (DCNN) model to handle the input sentences with varying length and induced a feature graph over the sentence. Such model is capable of explicitly capturing short and long-range relations. Kim [11] presented two simple CNN models with little hyper-parameter tuning for sentence-level classification.

3   Our Approach

In this study, we present a CNN-SVM combined model for sentiment analysis. The system framework is shown in Fig. 1. Firstly, CBOW model is applied to learn word embedding from a large collection of raw text. Secondly, a CNN model is applied to construct distributed sentence representations for input labeled data. Finally, the distributed sentence feature representations are used as the features for SVM classifier training by learning the probability distribution over labels.

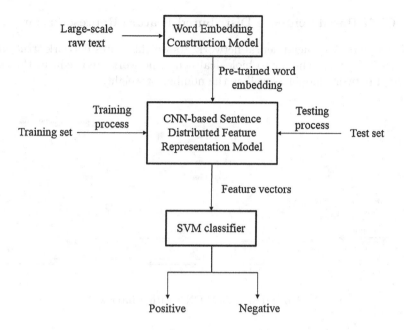

Fig. 1. Framework of CNN-SVM combined model.

3.1   Word Embedding Construction

Word embedding, wherein words are projected to a lower dimensional vector space via a hidden layer, are essentially feature learners that encodes semantic features of words in their dimensions. Mikolov et al. [13] introduced CBOW model to learn vector representations which captures syntactic and semantic word relationships from unlabeled text. The main idea is to predict a word by using its surrounding words in a context. Each word is mapped to a unique vector which is represented by a column in a matrix. The position of the word in the vocabulary is the index of the word in the matrix. The input layer of this model is the embeddings of surrounding words while the hidden layer is the concatenation of the word embedding which is used as features for predicting the target word in a sentence. Formally, given a sequence of words which are belong to a sentence in training data $w_1, w_2, \ldots, w_n$, the target of this model is to maximize the average log probability:

$$\frac{1}{n} \sum_{t=k}^{n-k} \log p(w_t | w_{t-k}, \ldots, w_{t+k}) \tag{1}$$

When the training converges, the words with similar semantic meanings are likewise close in the lower dimensional vector space.

3.2 CNN-Based Sentence Distributed Feature Representation

CNN is a kind of artificial neural network. Its weight shares network structures. It makes CNN more similar to biological neural network, and reduces the complexity of network model as well as the number of weight.

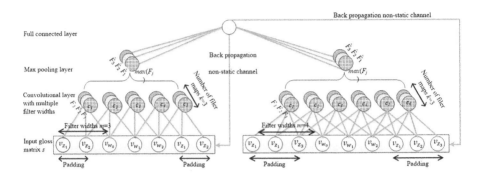

Fig. 2. A one-dimensional CNN with 2 filter widths.

As shown in Fig. 2, The one-dimensional CNN proposed by [11][1] is used for constructing distributed sentence feature representations. It includes four layers, namely input layer, convolutional layer, pooling layer, and fully connected layer. Let $w_i \in \mathbb{R}^d$ be the d-dimensional word embedding which is the i-th word in the sentence. A sentence s of length n (padded when the sentence length is under n) is represented as

$$s = [v_{z_1}, \cdots, v_{z_{m-1}}, v_{w_1}, \cdots, v_{w_n}, v_{z_1}, \cdots, v_{z_{m-1}}] \tag{2}$$

The vector obtained by concatenation is regarded as the input of the CNN model. A convolution operation involves a filter $w \in \mathbb{R}^{hk}$, which is applied to a window of h words to produce a new feature. The new feature is generated as follows:

$$c_i = f(w \cdot s_{i:i+h-1} + b) \tag{3}$$

where w and b are parameters of CNN. f is a non-linear function. $s_{i:i+m-1}$ refers to the i-th to the $(i + m - 1)$-th column of s. All of the possible windows of words in the sentence are applied to produce a feature map in the convolutional layer.

After convolution operation, a max-over-time pooling operation is applied to the feature map. The pooling operation takes the maximum value $\hat{c}$ in the feature map c and takes it as the feature corresponding to the particular filter. This operation constitutes the pooling layer and gets an m-dimension feature vector where m is the number of filters. The CNN model uses multiple filters with varying window sizes. These features are then transmitted to the last layer, namely fully connected layer, whose output is the probability distribution over labels.

---

[1] https://github.com/yoonkim/CNN_sentence.

The pre-trained word embeddings are fine-tuned via back propagation in the training processing of CNN model. Fine-tuning allows them to learn more meaningful representations of words. If the words do not appear in the pre-trained word embedding, they are initialized randomly. The vectors in the fully connected layer of CNN are regarded as the distributed sentence feature representations, and then these sentence representations are regarded as feature vectors in a SVM classifier.

3.3   CNN-based SVMs Classifier

SVM is a kind of supervised machine learning model for classification and regression analysis. Given a set of training examples, each marked for belonging to one of two categories, the SVM training algorithm builds a model that assigns new examples into one category or the other. Thus, SVM is a non-probabilistic binary linear classifier.

The CNN model for sentence classification, which is proposed by Kim [11], used a fully connected softmax layer as the classification layer. However, this classification layer is too simple to sentiment classification task. Fortunately, the output values of CNN pooling layer can be regarded as feature vectors for the input sentence. They may used as the input of other classifiers.

In this paper, we propose a CNN-based SVM classifier which treats CNN as the automatic feature learner and SVM as the sentiment classifier. The outputs of CNN, the distributed feature representations for the input sentences, are regarded as features in SVM. The SVM classifier is trained by using sentiment labeled sentences. When this model is applied to sentiment classification, the input sentences are transferred to distributed feature representations and then fed to SVM classifier for classification. Such a combined model is expected to combine the advantages of CNN and SVM.

4   Evaluation and Discussion

4.1   Experiment Settings

The dataset of NLPCC2014 Sentiment Classification with Deep Learning Technology Task (NLPCC-SCDL) is adopted to evaluate the proposed combined sentiment classification model. The NLPCC-SCDL task is designed to evaluate the sentiment analysis models based on deep learning. This dataset includes Chinese and English product reviews (Chinese side and English side for short, respectively) from multiple domains including book, DVDs and electronics. The statistics of NLPCC-SCDL are given is Table 1.

As for the evaluation metric, the ones adopted in NLPCC-SCDL evaluation are used here for fair evaluation and comparison. The metric is based on precision (P), recall (R), and $F1$ measure.

Let TP be the number of correctly classified positive samples, FP be the number of the falsely classified negative samples, TN be the number of the correctly classified negative samples, and FN be the number of the falsely classified positive samples. The metrics are defined as follows:

Table 1. The statistics of NLPCC-SCDL dataset

	Training set		Test set	
	Positive	Negative	Positive	Negative
English dataset	5000	5000	1250	1250
Chinese dataset	5000	5000	1250	1250

$$P_{pos} = \frac{TP}{TP + FP} \qquad P_{neg} = \frac{TN}{TN + FN}$$
$$R_{pos} = \frac{TP}{TP + FN} \qquad R_{neg} = \frac{TN}{TN + FP} \qquad (4)$$
$$F1_{pos} = \frac{2 \times P_{pos} \times R_{pos}}{P_{pos} + R_{pos}} \qquad F1_{neg} = \frac{2 \times P_{neg} \times R_{neg}}{P_{neg} + R_{neg}}$$

where the *pos* subscript refer to positive class and *neg* subscript refer to negative class.

Furthermore, we use classification accuracy (*Acc*) in the training process because single metric in the training process is helpful to parameter optimization in our proposed model.

$$Acc = \frac{TP + TN}{TP + FP + TN + FN} \qquad (5)$$

For pre-trained word embeddings, in the Chinese experiment, we trained the CBOW model by 50 million unlabeled Chinese sentences and finally obtained 399,059 word embeddings. In the English experiment, we use the publicly available vectors trained on 100 billion words from Google News by the CBOW model. All of the word embeddings have dimensionality of 300.

The CNN model proposed by Kim [11] is used. Meanwhile, we use SVM classifier with radial basis function in this experiment.

Four systems/algorithms are used as comparison systems. The first model is based on sentiment knowledge features, labeled as *Senti*. The second model, *TF*, refer to the model using word frequency features. The third one is a CNN-based model proposed by Kim which uses a fully connected layer as a classifier. The last one, labeled as $NLPCC - SCDLbest$, refers to the system which achieves the best performance in the NLPCC-SCDL evaluation [22]. For fair comparison, the adopted labeled dataset and evaluation metric are the same as in NLPCC-SCDL evaluation.

4.2   Experimental Results and Analysis

Training Stage. In the training process, the parameter optimization for CNN and SVM are conducted through closed testing on training data with 5-fold cross validation. Accuracy, *Acc*, is adopted in the training stage as the metric because the single metric is helpful to speed the parameter optimization. In CNN model,

Table 2. The parameter optimization results of CNN on training dataset with 5-fold cross validation (Chinese Side)

Filter \ Hidden unit	50	100	150	200
$[1, 2, 3]$	0.758	0.772	0.769	0.768
$[2, 3, 4]$	0.757	0.764	0.760	0.772
$[3, 4, 5]$	0.776	0.772	0.777	0.776
$[4, 5, 6]$	0.761	**0.778**	0.767	0.764
$[5, 6, 7]$	0.748	0.768	0.770	0.768
$[6, 7, 8]$	0.765	0.766	0.772	0.774

Table 3. The parameter optimization results of CNN on training dataset with 5-fold cross validation (English Side)

Filter \ Hidden unit	50	100	150	200
$[1, 2, 3]$	0.855	0.860	0.863	0.857
$[2, 3, 4]$	0.860	0.861	0.867	0.865
$[3, 4, 5]$	0.861	0.865	0.859	0.862
$[4, 5, 6]$	0.857	0.858	0.860	0.859
$[5, 6, 7]$	0.864	0.861	**0.868**	0.859
$[6, 7, 8]$	0.856	0.858	0.859	0.865

we adjust the parameter *Filter* and *Hidden unit* which *Filter* is the window sizes of filter and *Hidden unit* is the number of each window size filter. The product of *Filter* and *Hidden unit* is the size of feature vector which is the output of CNN pooling layer. The parameter optimization results of CNN are shown in Tables 2 and 3 on Chinese side and English side, respectively.

Table 2 shows that the classification accuracy on the Chinese dataset achieves the highest value of 0.778 when *Filter* is set to [4, 5, 6] and *Hidden unit* is set to 100. The highest accuracy value of 0.868 on the English dataset is achieved when *Filter* is set to [5, 6, 7] and *Hidden unit* is set to 150. Since the size of *Filter* and *Hidden unit* determine the time complexity of CNN model, these two parameters are selected within the appropriate range. Hence, in testing process, we let the parameter *Filter* be [4, 5, 6], *Hidden unit* be 100 and *Filter* be [5, 6, 7], *Hidden unit* be 150 of CNN for Chinese and English side, respectively.

In the process of parameters adjustment for SVM, the grid searching range of each parameter is: $\sigma = [10^{-3}, 10^{-2}, \ldots, 10^{2}, 10^{3}]$ and $C = [10^{-3}, 10^{-2}, \ldots, 10^{2}, 10^{3}]$. We tried 49 combinations on Chinese and English training dataset. The highest accuracy is achieved when C is 0.1 and *Gamma* is 0.01 for Chinese dataset, C is 0.01 and *Gamma* is 0.1 for English dataset, respectively.

Table 4 gives the closed testing results by our proposed model and the four comparison models.

It is observed that our proposed model achieves the highest classification accuracy on both Chinese side and English side. The 0.009 and 0.025 accuracy

Table 4. Classification accuracy by different models on training datasets

Dataset	Senti	TF	NLPCC2014 Best	CNN-based Model	Our model
Chinese	0.663	0.700	0.714	0.769	**0.778**
English	0.756	0.814	0.792	0.843	**0.868**

Table 5. Performance on NLPCC-SCDL testing dataset

Dataset	Model	Positive			Negative		
		P	R	F1	P	R	F1
Chinese dataset	NLPCC2014 best	0.758	0.789	0.773	0.780	0.748	0.764
	CNN-based model	0.759	0.796	0.777	0.781	0.749	0.764
	Our model	**0.766**	**0.806**	**0.785**	**0.795**	**0.754**	**0.774**
English dataset	NLPCC2014 best	0.856	0.866	0.861	0.864	0.855	0.860
	CNN-based model	0.871	0.860	0.865	0.860	0.873	0.866
	Our model	**0.890**	**0.886**	**0.888**	**0.886**	**0.891**	**0.889**

improvement on Chinese and English from the CNN-based model using fully connected layer are obtained, respectively. This shows the contribution of CNN and SVM combined model. Meanwhile, the achieved accuracy is obviously higher than the ones achieved by the $NLPCC - SCDLbest$, $Senti$ and TF. It is also observed that the achieved accuracies on English side are higher than Chinese side by all of the five tested systems. These results indicate that the sentiment analysis of Chinese is more complex and difficult.

Testing Process. In the testing stage, we use the optimized parameters of CNN and SVM. The final adopted parameters in testing stage are: $Filter$ is [4, 5, 6], $Hidden\ unit$ is 100, C is 0.1 and $Gamma$ is 0.01 for Chinese side and $Filter$ is [5, 6, 7], $Hidden\ unit$ is 150, C is 0.01 and $Gamma$ is 0.1 for English side. For the comparison CNN-based model, the same parameters are utilized. The achieved performances on Chinese and English dataset by different models are listed in Table 5. Noted that, the public evaluation metrics adopted in NLPCC-SCDL are utilized here.

It is observed that our proposed CNN and SVM combined model outperforms the CNN-based model and the NLPCC-SCDL best system. Compared to the CNN-based model, the $F1$ values on positive and negative categories by our model are increased for 0.8 % and 1.0 %, 2.3 % and 2.3 % on Chinese and English datasets, respectively. This result clearly shows that contribution of our CNN-SVM combined model by appending a SVM classifier.

Compared to the NLPCC-SCDL best system, our combined model obtains obvious improvement. The increments of $F1$ values on positive and negative are 1.2 % and 1.0 % for Chinese dataset, 2.7 % and 2.9 % for English dataset, respectively. The achieved performances are the highest one on NLPCC-SCDL dataset, based on our knowledge. These results show the effectiveness of our proposed model for sentiment analysis.

5 Conclusion

In this paper, we propose a CNN-SVM combined model for sentiment classification. This model treats CNN as a feature learner which automatically learns the feature representations for the input sentences. It is shown helpful to improve the feature representation and feature learning. Moreover, the combined model takes the advantages of SVM classifier on classification efficiency in order to improve the classification capability of CNN. The experiment results on the NLPCC2014 Sentiment Classification with Deep Learning Technology Task dataset show that our proposed combined model outperforms CNN model and the best system in the NLPCC 2014 evaluation, which shows the effectiveness of our proposed CNN-SVM combined model for sentiment classification.

Acknowledgements. This work was supported by the National Natural Science Foundation of China (No. 61370165, 61203378), National 863 Program of China 2015 AA015405, Natural Science Foundation of Guangdong Province S2013010014475, Shenzhen Development and Reform Commission Grant No.[2014]1507, and Shenzhen Peacock Plan Research Grant KQCX20140521144507925

References

1. Baccianella, S., Esuli, A., Sebastiani, F.: Sentiwordnet 3.0: an enhanced lexical resource for sentiment analysis and opinion mining. In: Proceedings of the International Conference on Language Resources and Evaluation, LREC 2010, 17–23 May 2010, Valletta, Malta (2010)
2. Bengio, Y., Ducharme, R., Vincent, P.: A neural probabilistic language model. J. Mach. Learn. Res. **3**, 1137–1155 (2003)
3. Collobert, R., Weston, J.: A unified architecture for natural language processing: deep neural networks with multitask learning. In: Proceedings of the 25th International Conference on Machine Learning (ICML), pp. 160–167. ACM (2008)
4. Collobert, R., Weston, J., Bottou, L., Karlen, M., Kavukcuoglu, K., Kuksa, P.: Natural language processing (almost) from scratch. J. Mach. Learn. Res. **12**, 2493–2537 (2011)
5. Graves, A., Mohamed, A.-R., Hinton, G.: Speech recognition with deep recurrent neural networks. In: Proceedings of the IEEE International Conference on Acoustics, Speech and Signal Processing (ICASSP), pp. 6645–6649. IEEE (2013)
6. Huang, M.,,Ye, B., Wang, Y., Chen, H., Cheng, J., Zhu, X.: New word detection for sentiment analysis. In: Proceedings of the 52nd Annual Meeting of the Association for Computational Linguistics (ACL), pp. 531–541. Association for Computational Linguistics, Baltimore,Maryland, June 2014

7. Joachims, T.: Text categorization with support vector machines: learning with many relevant features. In: Nédellec, C., Rouveirol, C. (eds.) ECML 1998. LNCS, vol. 1398. Springer, Heidelberg (1998)
8. Kalchbrenner, N., Grefenstette, E., Blunsom, P.: A convolutional neural network for modelling sentences. In: Proceedings of the 52nd Annual Meeting of the Association for Computational Linguistics, ACL 2014, 22–27 June 2014, Baltimore, MD, USA, vol. 1, pp. 655–665 (2014). Long Papers
9. Kalchbrenner, N., Grefenstette, E., Blunsom, P.: A convolutional neural network for modelling sentences. In: Proceedings of the 52nd Annual Meeting of the Association for Computational Linguistics (ACL), pp. 655–665. Association for Computational Linguistics, Baltimore, June 2014
10. Kim, S.M., Hovy, E.: Automatic detection of opinion bearing words and sentences. In: Proceedings of the International Joint Conference on Natural Language Processing (IJCNLP), pp. 61–66 (2005)
11. Kim, Y.: Convolutional neural networks for sentence classification. In: Proceedings of the 2014 Conference on Empirical Methods in Natural Language Processing (EMNLP), pp. 1746–1751, October 2014
12. Krizhevsky, A., Sutskever, I., Hinton, G.E.: Imagenet classification with deep convolutional neural networks. In: Advances in Neural Information Processing Systems 25: 26th Annual Conference on Neural Information Processing Systems 2012. Proceedings of a meeting held 3–6 December 2012, Lake Tahoe, Nevada, USA, pp. 1106–1114 (2012)
13. Mikolov, T., Sutskever, I., Chen, K., Corrado, G.S., Dean, J.: Distributed representations of words and phrases and their compositionality. In: Advances in Neural Information Processing Systems 26: 27th Annual Conference on Neural Information Processing Systems 2013. Proceedings of a meeting held 5–8 December 2013, Lake Tahoe, Nevada, USA, pp. 3111–3119 (2013)
14. Pang, B., Lee, L.: A sentimental education: sentiment analysis using subjectivity summarization based on minimum cuts. In: Proceedings of the 42nd Annual Meeting of the Association for Computational Linguistics (ACL), 21–26 July 2004, Barcelona, Spain, pp. 271–278 (2004)
15. Pang, B., Lee, L.: Seeing stars: exploiting class relationships for sentiment categorization with respect to rating scales. In: Proceedings of the 43rd Annual Meeting on Association for Computational Linguistics (ACL), pp. 115–124. Association for Computational Linguistics (2005)
16. Pang, B., Lee, L., Vaithyanathan, S.: Thumbs up?: sentiment classification using machine learning techniques. In: Proceedings of the ACL-02 Conference on Empirical Methods in Natural Language Processing, vol. 10, pp. 79–86 (2002)
17. Salvetti, F., Lewis, S., Reichenbach, C.: Automatic opinion polarity classification of movie. Colorado Res. Linguist. **17**, 2 (2004)
18. Shen, Y., He, X., Gao, J., Deng, L., Mesnil, G.: Learning semantic representations using convolutional neural networks for web search. In: Proceedings of the Companion Publication of the 23rd International Conference on World Wide Web Companion, pp. 373–374. International World Wide Web Conferences Steering Committee (2014)
19. Tang, D., Wei, F., Qin, B., Zhou, M., Liu, T.: Building large-scale twitter-specific sentiment lexicon: a representation learning approach. In: Proceedings of the 25th International Conference on Computational Linguistics (COLING), pp. 172–182 (2014)

20. Tekiroğlu, S.S., Özbal, G., Strapparava, C.: A computational approach to generate a sensorial lexicon. In: Proceedings of the 25th International Conference on Computational Linguistics (COLING), p. 114 (2014)
21. Tong, R.M.: An operational system for detecting and tracking opinions in on-line discussion. In: Working Notes of the ACM SIGIR 2001 Workshop on Operational Text Classification, vol. 1, p. 6 (2001)
22. Wang, Y., Li, Z., Liu, J., He, Z., Huang, Y., Li, D.: Word vector modeling for sentiment analysis of product reviews. In: Zong, C., Nie, J.-Y., Zhao, D., Feng, Y. (eds.) NLPCC 2014. CCIS, vol. 496, pp. 168–180. Springer, Heidelberg (2014)
23. Xu, L., Lin, H., Zhao, J.: Construction and analysis of emotional corpus. J. Chin. Inf. Process. **22**(1), 116–122 (2008)
24. Yang, Y., Liu, X.: A re-examination of text categorization methods. In: Proceedings of the 22nd Annual International ACM SIGIR Conference on Research and Development in Information Retrieval, pp. 42–49. ACM (1999)
25. Yih, W.-T., He, X., Meek, C.: Semantic parsing for single-relation question answering. In: Proceedings of the 53rd Annual Meeting on Association for Computational Linguistics (ACL) (2014)

A Novel Cross Modal Hashing Algorithm Based on Multi-modal Deep Learning

Wen Qu[1]([✉]), Daling Wang[1,2], Shi Feng[1,2], Yifei Zhang[1,2], and Ge Yu[1,2]

[1] School of Information Science and Engineering,
Northeastern University, Shenyang, China
`quwen@research.neu.edu.cn`
[2] Key Laboratory of Medical Image Computing,
Northeastern University, Ministry of Education, Shenyang, China
{`wangdaling,fengshi,zhangyifei,yuge`}`@ise.neu.edu.cn`

Abstract. With the popularity of multi-modal data on Web, cross media retrieval has become a hot research topic. Existing cross modal hash methods assume that there is a latent space shared by multi-modal features, and embed the heterogeneous data into a joint abstraction space by linear projections. However, these approaches are sensitive to the noise of data, and unable to make use of unlabelled data and multi-modal data with missing values in the real-world applications. To address these challenges, in this paper, we propose a novel Multi-modal Deep Learning based Hashing (MDLH) algorithm. In particular, MDLH adopts deep neural network to encode heterogeneous features into a compact common representation and learn the hash functions based on the common representation. The parameters of the whole model are fine-tuned in supervised training stage. Experiments on two standard datasets show that our method achieves more effective results than other methods in cross modal retrieval.

1 Introduction

As the popularity of social media in the Web 2.0, the amount of multi-modal data increases dramatically in recent years. For example, photos are usually associated with captions and tags, videos contain visual and audio signals, and tweets often consist of text, images and videos. At the same time, when users acquire and search through the Internet, they also want to get a comprehensive result consisting of multiple media types. The traditional information retrieval system only uses text as query input, so most information systems provide the image and video retrieval based on text queries. With the rapid development of the mobile equipment such as telephone and flat computer, users may perform queries using image, audio and videos other than text. There is an emerging need to retrieve and search similar or relevant data entities from multiple modals. To make the

---

W. Qu—This work is supported by the National Natural Science Foundation of China under Grant No. 61402091, 61370074 and the Fundamental Research Funds for the Central Universities of China under Grant No. N140404012.

© Springer Science+Business Media Singapore 2015
X. Zhang et al. (Eds.): SMP 2015, CCIS 568, pp. 156–167, 2015.
DOI: 10.1007/978-981-10-0080-5_14

system possible for handling large amount of multimedia data, hashing based methods have attracted increasing attentions due to the advantages in reducing both the computational cost and storage. A lot of work extended uni-modal hashing into multi-modal setting [23]. Cross modal hashing maps data of different modalities into the hamming space, in which the distance of similar objects to be small. In the hamming space, all data are represented as hash codes and can be searched quickly even for the databases with millions of data. Most previous cross modal hashing methods follow the assumption that multi-modal data used for training are available in all the multiple modals and contain the same 'semantic object'. So these works can not make use of unlabeled data or multi-modal data with missing values. In realistic applications, the data in the Internet is typically very noisy and may have missing modals. For example, the image and text of a tweet may contain different semantics at all. Furthermore, given a system supporting cross modal retrieval including text, image and audio, if the data generated by users only contain text and image, they cannot be used for modeling relationship among the three modals. Most previous works represent multi-modal data through clustering [25], dictionary learning [22], which build the corresponding maps pair-wise. When a new modal is added to the system, the relationship of the new modal with each existing modal has to be learned again. To address these problems, in this paper we propose a Multi-modal Deep Learning based Hashing (MDLH) algorithm, which learn the common feature space of different modalities using deep neural network. The multi-modal deep learning can learn compact and robust 'semantic' representation of multi-modal data, which is able to handle the noise and the missing modals of the data. The experiments on two realistic datasets show that the proposed method can realize cross modal hashing effectively. The rest of the paper is organized as follows: In Sect. 2, we review the related work. Section 3 elaborates the method proposed in this paper. In Sect. 4, we demonstrate the use of our approach for cross modal retrieval and the experimental results. Finally, we conclude the work in Sect. 5.

2   Related Work

The work involves with cross modal hashing and multi-modal deep learning, which will be reviewed in the following subsections.

2.1   Cross Modal Hashing

Hashing index can be categories into uni-modal hashing, multi-modal hashing, and cross modal hashing. In the work about uni-modal hashing, the most well known methods are local sensitive hashing [5] and spectral hashing [20]. Multi-modal hashing compares the multi-modal features of data, and returns the search results of each modal. For example, when retrieving an image according to multi-modal (color, SIFT, BOW) descriptors, the multi-modal hashing projects each feature into the hamming space and combines the multiple results together. Cross modal hashing focuses on analyzing the relationship between modalities and provides cross modal query. For example, given the color feature of an image as the input, the system returns the results according to SIFT descriptor. Here the modal means feature or media type. So cross modal means cross feature or cross media. The existing uni-modal data hashing includes two steps: First, project the original data into low-dimensional space. Then, quantize the new representation into hash codes. Under the unsupervised situation, many embedding methods have been proposed, such as random projection [5], spectral decomposition [20]. Similarly, multi-modal data hashing includes two steps with more restrictions. Bronstein [3] proposed the first cross modal hashing model CMSSH. Given two modals, CMSSH learned two groups of hash functions that made the similar data (in different modals) have smaller distance in the hamming space while dissimilar data (in different modals) have larger distance in the hamming space. CMSSH kept the relationship between data in different modals but ignored the similarity in same modal. Kumar [10] extended the spectral hashing into multi-modal setting and proposed CVH, which minimized the distance of similar data both in the same modal and the different modals. MLBE [23] used probability generative model to represent the data, and the latent factors learned were used as the hash codes. There is no independent restrict of hash codes so the hash codes may have high redundancy. Yu [22] adopted dictionary learning to represent data in different modals, and learned the hash function based on sparse codes. The dictionaries of different modals were connected through the coupled dictionary space. IMVH [8] kept both intra similarity and inter similarity of the data. Song et al. [17] proposed Inter-Media Hashing which used a set of corresponding image and text as the inter media to learn the relations of multiple modals.

2.2   Multi-modal Deep Learning

Deep learning builds a layer network structure to simulate the human brain, and learn representations for data from bottom to up. Each layer of the network corresponds to a representation. Recently, deep learning is widely used in many applications and achieves impressive results, including speech recognition [6], face recognition, image classification [9] and object recognition etc. The representative deep learning including Deep Belief Networks [1], Auto Encoder, Stacked Denoising Autoencoder, Deep Boltzmann Machine and Deep Energy Model. Ngiam [12] used DBM to learn the cross modal representation of video and audio data, and rebuilt the data of missed modality. Srivastava [18] proposed a Deep Belief Network to learn representation of the multi-modal data. Sohn [16] proposed an improved multi-modal deep learning model. These works focus on solving the data rebuilding problem when part of the modality is missing. Our work focuses on learning the relations between different modals and proposing a semantic and common representation of multi-modal data. The work most similar to ours is Wu [21], in which the deep learning is used to learn the optimal combination of different modalities. Different from their work, we focus on learning a common representation of multi-modal data using deep neural network.

3 The Multi-modal Deep Learning Based Hashing Algorithm Methodology

In this section, we present the MDLH algorithm in detail. Figure 1 is the framework of our method. First, the multi-modal features of multi-modal data are extracted as inputs. Then, we use multi-modal deep learning method to learn the common representation for them. Finally, the hash function of each modality is used to map the data into the hamming space. In the following, the notations and problem formulation are introduced first. Then, we give the model of the multi-modal deep learning, followed by hashing function learning.

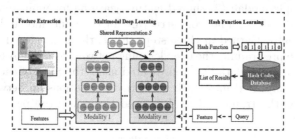

Fig. 1. Framework of the multi-modal deep learning hashing

3.1 Notations and Problem Definition

Given a set of multi-modal data $O = O^1, ..., O^p, ..., O^M (p = 1...M)$ consist of N data from M modalities, where O^p is the dataset in the p-th modal and o_i^p is the i-th datum in O^p. We use X^p to represent the features of the p-th modal, and D_p is the dimension of the feature space. Denoted the shared representation of multi-modal data is S, the projections are defined as:

$$f^p : X^p \to S^p \tag{1}$$

then, the data are mapped into hamming space using a linear projection:

$$g^p : S^p \to H^p \tag{2}$$

The main idea of learning the hash functions goes as follows. Data of each individual modal are firstly converted into the representations for single modal, denoted as B^p, which preserves the intra similarity. Data of all modals represented by B^p are then mapped into a common space S^p where the inter-similarity is preserved to generate hash functions. Finally, values of hash functions are binarized into hamming space. Given a set of multi-modal data O and the training dataset $T = (x_i^{m_i}, x_j^{m_j})^k, k = 1, ..., K$, where $x_i^{m_i} \epsilon O^{m_i}$, $x_j^{m_j} \epsilon O^{m_j}$ are the features of $o_i^{m_i}$ and $o_i^{m_j}$ separately. $L_{ij} = 1$ if two data x_j and x_j belong to the same category otherwise $L_{ij} = -1$. The distance of the two data in the shared representation is defined as:

$$d(x_i^{m_i}, x_j^{m_j}) = \|s_i^{m_i} - s_j^{m_j}\|_F^2 \tag{3}$$

We formulate the problem to the following optimal problem with the object function:

$$\min_{f} \sum_{k=1}^{K} L_{ij} d(x_i^{m_i}, x_j^{m_j}).$$ (4)

3.2 Multi-modal Deep Learning

In this section, we describe the multi-modal feature learning model for the task of shared representation learning, where the inputs are the features of each modal. The multi-modal deep learning consists of two components: (1) feature learning for each single modal; (2) shared feature learning for multi-modal features. Figure 2 is the deep neural network structure for the multi-modal deep learning. The whole model is learned in three steps: First, the unlabeled data U of each modal is used to pre-training the deep learning network using SDA (seeing 3.2.1). Then, the multi-modal data O is represented using the SDA of each modal and the outputs are inputted into RMB to learn the relationship between multiple modals. Finally, the training data T is used to update the parameters of the model.

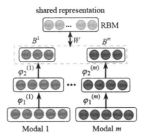

Fig. 2. Multi-modal deep learning model

Feature Learning for Single Modal. The SDA (Stacked Denoising Autoencoder [19]) is adopted to pre-training the network, which adds noises into training data based on autoencoder. Figure 3 is the process of a SDA. First, we construct the noisy version of x through a stochastic mapping. Then the noisy version x' is mapped through AE to a hidden representation $y = \varphi(x')$, where y is used to reconstruct a clean version of x by $z = \psi(y)$. Several DAs are stacked to build a layer structure, where the output of the bottom layer is the input to the higher layer. Once the encoding function is learned, encoding function is not needed anymore. We use a non-linear one-layer neural network as the unit of SDA, where the encode function is:

$$y = \varphi(x) = sigmoid(Qx + r)$$ (5)

and decoding function is:

$$z = \psi(y) = sigmoid(Q'y + t)$$ (6)

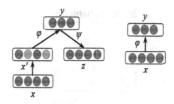

Fig. 3. Denoising Autoencoder

Multi-modal Feature Learning. After learned the representation of each modality, we use RBM (Restricted Boltz-mann Machine [15]) to model the relations between different modals and learn the shared representation of them. A Restricted Boltzmann Machine is an undirected graphical modal with stochastic visible unit v and stochastic hidden unit h. Each visible unit connects to each hidden unit, but no connections within hidden variables or visible variables. The structure of the model is shown in Fig. 4. The model defines the following energy function E:

$$E(v, h; \theta) = -a^T v - b^T h - v^T W h \tag{7}$$

where $\theta = \{a, b, W\}$ are the model parameters. The joint distribution over the visible and hidden units is defined by:

$$p(v, h; \theta) = \frac{1}{Z(\theta)} exp(-E(v, h; \theta)) \tag{8}$$

where $Z(\theta)$ is a constant for normalization. The j-th hidden node is set to 1 with probability:

$$p(h_j|v) = sigmoid(\frac{1}{\sigma^2}(b_j + W_j^T v)) \tag{9}$$

We minimize the loss function between reconstructed data using the model and original data, and learn the parameter following [7]. After obtaining the shared representation s by the multi-modal deep learning model, we can compute the derivation of the objection function with respect to $s_i^{m_i}$ and $s_j^{m_j}$ as follows:

$$\frac{\partial J}{\partial s_i^{m_i}} = 2 \sum_{k=1}^{K} L_{ij}(s_i^{m_i} - s_j^{m_j}) \tag{10}$$

$$\frac{\partial J}{\partial s_j^{m_j}} = 2 \sum_{k=1}^{K} L_{ij}(s_j^{m_j} - s_i^{m_i}) \tag{11}$$

Then, we used online gradient descent [26] to update the parameter of the last layer by:

$$W \leftarrow W - \eta \frac{J}{W} \tag{12}$$

$$b \leftarrow b - \eta \frac{J}{b} \tag{13}$$

where the derivative are computed as follows:

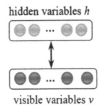

hidden variables h

visible variables v

Fig. 4. Restricted Boltzmann Machine

$$\frac{\partial J}{\partial W} = \frac{\partial J}{\partial s_i^{m_i}} \frac{\partial s_i^{m_i}}{\partial W} + \frac{\partial J}{\partial s_j^{m_i}} \frac{\partial s_j^{m_i}}{\partial W} \tag{14}$$

$$\frac{\partial J}{\partial b} = \frac{\partial J}{\partial s_i^{m_i}} \frac{\partial s_i^{m_i}}{\partial b} + \frac{\partial J}{\partial s_j^{m_i}} \frac{\partial s_j^{m_i}}{\partial b_j^m} \tag{15}$$

Finally, we adopt back propagation [14] to update the parameter in the other layers of the network.

3.3 Hashing Function Learning

Let us denote the shared representation for a data is s, the linear transformation to hash code is:

$$g(s) = sign(P^T s) \tag{16}$$

where P is the projection matrix, s is the shared representation of data. Denote $S = [S^1, ..., S^M]$ as the representation for all dataset. Since our representation S is sparse, we follow the method in [22] to learn the projection matrix P by:

$$P = \sqrt{MN} \Lambda^{-\frac{1}{2}} V \Sigma^{-\frac{1}{2}} \tag{17}$$

Algorithm 1. Multi-modal deep learning based cross modal hashing

---

Input: multi-modal data O, training data U, T
Output: projection f^p, projection g^p

1. **for** $m = 1 : M$
2. pretrainning the SDA for modality m
3. **end**
4. pretrainning the RBM
5. **do**
6. **for** $k = 1 : K$
7. $(x_i, x_j) \leftarrow T$
8. update the parameter W, b
9. update the parameter in the lower layer using back propagation
10. **end**
11. **Untile** object function convergence
12. Compute P using equation(17)

---

where M and N are the number of modal and the multi-modal data, V and Σ are the c largest eigenvalue and corresponding eigenvector of the matrix $\Lambda^{-\frac{1}{2}} SS^T \Lambda^{-\frac{1}{2}}$ with $\Lambda = diag(S)$. Algorithm 1 summarizes the multi-modal deep learning based cross modal hashing. Given a new data, the hash code is generated by two steps: First, extract the feature of the data and use the multi-modal deep learning to represent the data. Then, use the linear project function g to compute the hash code of the data.

4   Experiments

We evaluate our method on two real-world datasets for cross modal similarity search and analyse the results. In detail, the datasets consist of text and images, and we use text as query to search similar images and image as query to search similar texts. First, we introduce the dataset and the setting of the experiments. Then we will show the results and compare the results with other methods.

4.1   Data Sets and Settings

Two datasets are used in our experiment: Wikipedia-Picture of the Day and NUS-WIDE. All of them include two modals (pictures and text). Wikipedia [13] includes 2866 multimedia documents collected from Wikipedia website, in which each document includes one picture and at least 70 words. The dataset provides the topic probability of each text on 10 categories (computed using LDA [2]). Existing experiments used the topic probability as text features, which is too sparse to be a suitable input to deep learning. So we extract the vector space modal of each text as the feature. The feature of images use SIFT descriptor [11] based on bag-of-visual word model, which quantizes the descriptors into 1,000 dimensional vectors. The NUS-WIDE dataset is a real-world image dataset collected by Lab for Media Search in National University of Singapore [4]. It includes 81 categories and 269,648 images. Each image corresponds to multiple tags, and each image-text pair is annotated by at least one category. The image is represented by 1000-dimensional bag-of-visual word of SIFT descriptors. And the text corresponding to the image is represented by a 1000-dimensional vector of tags.

4.2   Evaluation Metric

We use mean Average Precision [23] as the evaluation metric for effectiveness in our experiment. The evaluation metric has been widely used in literatures [23,24]. The mAP evaluates the performance of similarity search, which the larger value indicates better performance and the similar results have high ranks. Given a query and R retrieved instances, the average precision is defined as:

$$AP = \frac{1}{L} \sum_{r=1}^{R} P(r)\delta(r)$$ (18)

where L is the number of relevance instances in the result. $P(r)$ is the accuracy of top r instances. $\delta(r)$ is indicator function, which equals to 1 if the r-th instance is relevant to the query or 0 otherwise. The mAP is the mean of all the APs and we set $R = 100$ in our experiments.

4.3   Compared Methods

We compared our method with other four cross modal hash methods. They are CMSSH, CVH, LSSH and IMVH. CMSSH [3] constructed two groups of linear hash function to keep similarity relationship between different modalities. CVH [10] extended uni-modality spectral hashing to multi-modal, and kept the similarity relationship between different modal and in the same modal. LSSH [24] adopted matrix factorization and sparse coding to map text and image into the latent factor space separately. IMVH [8] kept the interior and exterior similarity, and added the distinctive into data belongs to different categories. CMSSH and CVH generate different hash codes for different modals, but they make sure that the hash codes in the same modal have the same length.

4.4   Results

Results on Wiki Dataset. We select 90 % of the dataset as training data, 5 % as unlabeled data and the rest as the query set for MDLH. Other methods use the 95 % of the dataset (training data and unlabeled data for MDLH) as training data and the rest as the query set. The mAP of our method and compared methods on Wiki dataset are shown as Table 1. We can observe that MDLH outperforms most of the methods on two cross modal similarity search tasks. The results of existing work reported better performance on task 'Text query Image' than task 'Image query Text', because they used topics rather than words as the text feature so the text queries are represented as the 10 topic, which simplify the research problem. Furthermore, we report the Top-N precision curve of results on Wiki dataset in Fig. 5, which reflects the change of precision with respect to the number of retrieved instances.

Results on NUSE-WIDE Dataset. Some categories in NUSE-WIDE are scarce, so we select 8 categories that contain more instances than the other categories. We select 90 % of the dataset as the training data, 5 % as unlabeled data, and 5 % as the query data for MDLH. The mAP of all the methods on NUW-WIDE dataset is shown in Table 2. The performance of all methods increased to some degree on NUS-WIDE dataset.

Results on Noised Dataset. To evaluate the robustness to noise of each method, we add noises into Wiki and NUS-WIDE datasets separately, and compare the performance on the noise dataset. For Wiki and NUS-WIDE dataset, we select a category randomly as the source of noise separately. Some pictures and words from them are selected randomly as noise adding to the rest of the data.

Table 1. The mAP of different methods on Wiki dataset

Task	Method	Hash code length		
		16	32	64
Image query Text	CMSSH	0.3183	0.3275	0.2750
	CVH	0.3140	0.3345	0.2760
	LSSH	0.3730	0.3940	0.3887
	IMVH	0.3812	0.3921	0.3879
	MDLH	**0.3919**	**0.3940**	**0.4030**
Text query Image	CMSSH	0.3321	0.3173	0.3147
	CVH	0.3005	0.3322	0.3107
	LSSH	0.3552	0.3559	0.3545
	IMVH	0.3642	0.3624	0.3644
	MDLH	**0.3840**	**0.3729**	**0.3604**

Table 2. The mAP of different methods on NUS-WIDE dataset

Task	Method	Hash code length		
		16	32	64
Image query Text	CMSSH	0.4405	0.4389	0.3934
	CVH	0.3756	0.3729	0.3619
	LSSH	0.4517	0.4437	0.4460
	IMVH	0.4520	0.4489	0.4446
	MDLH	**0.4526**	**0.4537**	**0.4555**
Text query Image	CMSSH	0.4113	0.3984	0.3722
	CVH	0.3805	0.3629	0.3899
	LSSH	0.4271	0.4178	0.4143
	IMVH	0.4189	0.4250	0.4130
	MDLH	**0.4496**	**0.4478**	**0.4485**

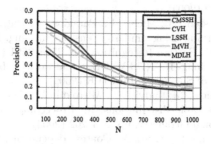

Fig. 5. Top-N precision of different methods on Wiki dataset

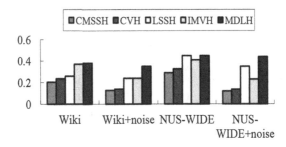

Fig. 6. The mAP of different methods with and without noise

In Wiki dataset, we select 2 % of the text and one picture as noise each time. In NUS-WIDE dataset, we select one tag as the noise. Figure 6 is the performance before and after adding noises. It shows that our method is robust to noise than other methods.

5   Conclusion

In this paper, we proposed a multi-modal deep learning based cross modal hash learning method. The multi-modal deep learning is used to model the relationship between multiple heterogeneous data and learn a shared representation of the multi-modal data, which is robust to noise and easy to extend to multiple modals. The experiments on two realistic dataset show that our method representing the multi-modal features effectively. In future, we will focus on the multi-modal deep learning for media types such as audio, video.

References

1. Bengio, Y., Lamblin, P., Popovici, D., Larochelle, H.: Greedy layer-wise training of deep networks. In: NIPS, pp. 153–160. MIT Press (2006)
2. Blei, D., Ng, A., Jordan, M.: Latent dirichlet allocation. J. Mach. Learn. Res. (JMLR) **3**, 993–1022 (2003)
3. Bronstein, M., Bronstein, A., Michel, F., Paragios, N.: Data fusion through cross-modality metric learning using similarity-sensitive hashing. In: Proceedings of the Computer Vision and Pattern Recognition, pp. 3594–3601 (2010)
4. Chua, T., Tang, J., Hong, R.: NUS-WIDE: a real-world web image database from National University of Singapore. In: CIVR (2009)
5. Datar, M., Immorlica, N., Indyk, P., Mirrokni, V.: Locality-sensitive hashing scheme based on p-stable distributions. In: Proceedings of ACM Annual Symposium Computational Geometry, pp. 253–262 (2004)
6. Dahl, G., Yu, D., Deng, L., Acero, A.: Context-dependent pre-trained deep neural networks for large-vocabulary speech recognition. TASLP **20**(1), 30–42 (2012)
7. Hinton, G., Salakhutdinov, R.: Reducing the dimensionality of data with neural networks. Science **313**(5786), 504–507 (2006)
8. Hu, Y., Jin, Z., Ren, H., Cai, D., He, X.: Iterative multi-view hashing for cross media indexing. In: ACM Multimedia, pp. 527–536 (2014)

9. Krizhevsky, A., Sutskever, I., Hinton, G.: Imagenet classification with deep convolutional neural networks. In: NIPS, pp. 1106–1114 (2012)
10. Kumar, S., Udupa, R.: Learning hash functions for cross-view similarity search. In: International Joint Conference on Artificial Intelligence, pp. 1360–1365 (2011)
11. Lowe, D.: Distinctive image features from scale-invariant key points. Int. J. Comput. Vis. 60(2), 91–110 (2004)
12. Ngiam, J., Khosla, A., Kim, M., Nam, J., Lee, H., Ng, A.Y.: Multimodal deep learning. In: ICML, pp. 689–696 (2011)
13. Rasiwasia, N., Pereira, J., Coviello, E., Doyle, G., Lanckriet, G.R., Levy, R., Vasconcelos, N.: A new approach to cross-modal multimedia retrieval. In: ACM MM, pp. 251–260 (2010)
14. Rumelhart, D., Hinton, G., Williams, R.: Neurocomputing: Foundations of Research. MIT Press, Cambridge (1988)
15. Salakhutdinov, R., Hinton, G.: Deep boltzmann machines. In: AISTATS, vol. 5, pp. 448–455 (2009)
16. Sohn, K., Shang, W., Lee, H.: Improved multimodal deep learning with variation of information. In: INPS, pp. 2141–2149 (2014)
17. Song, J., Yang, Y., Yang,Y., Huang, Z., Shen, H.T.: Inter-media hashing for large-scale retrieval from heterogeneous data sources. In: SIGMOD, pp. 785–796 (2013)
18. Srivastava, N., Salakhutdinov, R.: Multimodal learning with deep boltzmann machines. In: NIPS, pp. 2231–2239 (2012)
19. Vincent, P., Larochelle, H., Lajoie, I., Bengio, Y., Manzagol, P.-A.: Stacked denoising autoencoders: learning useful representations in a deep network with a local denoising criterion. JMLR 11, 3371–3408 (2010)
20. Weiss, Y., Torralba, A., Fergus, R.: Spectral hashing. In: Advances in Neural Information Processing Systems (2005)
21. Wu, P., Hoi, S., Xia, H., Zhao, P., Wang, D., Miao, C.: Online multimodal deep similarity learning with application to image retrieval. In: ACM Multimedia, pp. 153–162 (2013)
22. Yu, Z., Wu, F., Yang, Y., Tian, Q., Luo, J., Zhuang, Y.: Discriminative coupled dictionary hashing for fast cross-media retrieval. In: SIGIR, pp. 395–404 (2014)
23. Zhen, Y., Yang, D.: A probabilistic model for multimodal hash function learning. In: SIGKDD, pp. 940–948 (2012)
24. Zhou, J., Ding, G., Guo, Y.: Latent semantic sparse hshing for cross-modal similarity search. In: SIGIR, pp. 415–424 (2014)
25. Zhu, X., Huang, Z., Shen, H., Zhao, X.: Linear cross-modal hashing for efficient multimedia search. In: ACM Multimedia, pp. 143–152 (2013)
26. Zinkevich, M.: Online convex programming and generalized infinitesimal gradientascent. In: ICML, pp. 928-936 (2003)

Predicting Who Will Retweet or Not
in Microblogs Network

Jun Zhou [1,2]([✉]), Zhen Zhang[3], Bing Wang[1], Yan Zhang[1], and Yonghong Yan[1]

[1] The Key Laboratory of Speech Acoustics and Content Understanding,
Institute of Acoustics, Chinese Academy of Sciences, Beijing, China
zhoujun@iie.ac.cn
[2] Institute of Information Engineering, Chinese Academy of Sciences, Beijing, China
[3] National Computer Network Emergency Response Technical Team/Coordination
Center of China, Beijing, China

Abstract. Retweeting often leads to fast and wide information propagation in microblogs. There has been research concerning predicting users' retweeting behavior. However, it lacks of in-depth investigations on key features that contribute to the prediction accuracy. In this paper, we systematically examined four types of features: followee (the user who posted tweets) features, follower features, tweet features and interaction features in terms of their contribution to retweeting prediction accuracy. We collected Sina Weibo data and ranked the features extracted from original data on importance by using an information gain method. There were twelve features showing significant association with the retweeting predictive accuracy through lots of experiments. Experimental results showed that using these twelve features, we could train a novel predictive model and achieved the predict accuracy up to 98 %.

Keywords: Retweet · Social network · Tweet · Information propagation

1 Introduction

Due to the data richness and public availability of various microblogging system, researchers have been using such systems to investigate how and why certain information triggers people's attention and relaying this information [1]. Sina Weibo is a microblogging service that provides its users with an information sharing platform, and retweeting is widely accepted as an information propagation form [2]. Researchers have been interested in using retweeting related data to study how information propagates in social networks.

Petrovic et al. [3] predicted if a tweet would be retweeted using a machine learning approach based on the passive-aggressive algorithm and Li et al. [4] proposed a novel approach to predict the retweet count of a tweet, not predicting if a tweet would be retweeted by a given follower. Lee et al. [5] addressed how to actively identify and engage the right strangers at the right time on social media to help effectively propagate intended information within a desired time frame,

© Springer Science+Business Media Singapore 2015
X. Zhang et al. (Eds.): SMP 2015, CCIS 568, pp. 168–175, 2015.
DOI: 10.1007/978-981-10-0080-5_15

but they didn't study retweeting behavior in depth. The properties of retweets, relating these to the friend-follower graph and to message propagation speeds, was preliminary studied [6]. Suh et al. [7] just examined 9 feature including url, hashtag, mention, follower, followee, days, status, favorite, retweet, building a predictive retweet model. A novel model was constructed to capture the three major properties of information diffusion: speed, scale, and range in the study carried out by Yang et al. [8], and they just predicted information diffusion roughly and didn't care who retweet it. Ota et al. [9] discovered Twitter users who retweeted many tweets by overlapping propagation paths of retweets. The influence of users in Twitter was analyzed, using three different measures including indegree, retweets, and mentions [10,11]. Bao et al. [12] found that the popularity of content was well reflected by the structural diversity of the early adopters, and they mostly considered tweet contents.

In this paper, we are interested in investigating such questions as: Given an original tweet from some one, how to predict who will retweet it among his/her followers? What are the features that contribute significantly to such a prediction? As discussed above, the novelty in this study is that this work is focused on not only tweet contents but also user behaviors. Besides, a comprehensive features in follower-followee relations are considered.

The rest of the paper is organized as follows: in Sect. 2, we describe the four types of features: followee features, follower features, tweet features and interaction features. In Sect. 3, we describe data set and data preprocessing, and present our experiments and analyse results. In Sect. 4, we present the conclusion of our research and future works.

2   Features

We divide the related features into four distinct categories: followee features, follower features, tweet features, and interaction features. The followee features refer to the properties related to the authors of original tweets. The follower features describe the properties of followers. The tweet features include various statistics about the tweet contents. The interaction features reflect interactive properties between followees and followers. The detailed components of each category are introduced in the following sections.

2.1   Followee and Follower Features

Both the followee and follower features indicate user properties. They share common properties. We use the following features for user features: number of followers, number of friends, number of statuses, number of favorites, number of bi-followers, url, verification, language, length of name, length of description, days, province, city, gender and user influence.

Next, we explain some features except that the features easy to understand. The number of statuses is the number of the tweets posted. Bi-followers indicates that the user A follows the user B and the user B also follows the user A.

Verification is used by SinaWeibo mostly to confirm the authenticity of celebrity accounts. Sina Weibo has three languages setting: simplified Chinese, traditional Chinese and English. The length of name/description is the count of characters contained. Days indicate the length of registration time. Province and city show where the user locates. In general, we regard a person who has a number of followers and a few friends as influential, so we define user influence:

$$UI = \frac{countOfFollowers}{countOfFriends + 1}. \tag{1}$$

2.2 Tweet Features

We use the following features for tweet contents: number of hashtags, number of URLs, number of pictures, number of @ and length of the tweet. The length of the tweet indicates the count of actual words in the tweet.

2.3 Interaction Features

There is a type of retweeting involved in conversations. When a follower comments on a tweet, she/he can use retweeting with a comment. Close friends often have more interactions than normal acquaintances do. We hence take relations between followers and followees into consideration for reflecting such a property in this study. Firstly, if two users (follower-followee pair) share more friends, they are likely to form more conversations. Second, if a followee also follows some his/her followers, intuitively such relations are stronger than one-sided followings, then it is more likely that retweets happen among them. Since we can extract user locations from weibo log, it is possible to define the distance between followees and their followers. Being geographically close implies that there are activities that two users can both attend, which in turn promotes interactions. Therefore, in this study, distance is considered as a feature in the predication task. If they don't live in the same province, we define the distance to be far; If they live in the same province, but in different cities, we define the distance to be medium; If they live in the same city, we define the distance to be near. As we know weibo users' genders, we use this information to add to follower/followee relations with one of the following gender properties: 'mm' (a followee's gender is male and his/her one follower's is also male), 'mf', 'fm' and 'ff'. As described above, we know some user features whose value are numeric, so the ratio of the two numbers forms a new feature such as the ratio of the count of their followers.

3 Experiments

We compare different performance of five popular classifiers: BayesNet, IBk (KNN), Logistic (Logistic Regression), RandomForest and J48 (C4.5). We use WEKA [13] implementation of these algorithms and train these classifiers to predict the probability of a person to retweet and classify a follower as a retweeter or non-retweeter.

3.1 Data Set

Our data which collected from Sina Weibo contain the following information. For a user, we can get his/her name (screen name), location, description, gender, registration date, number of followers and friends and so on. For a tweet, we can get the author of the tweet, time when the user posted the tweet, the content of the tweet, and whether there are pictures contained in the tweet. If the tweet contains a retweet, we can get same information as above. If the tweet doesn't contain a retweet, it is original. Here the tweets we experiment are all original.

3.2 Preprocessing

For preprocessing, we first extract the retweets from our data. Second, using the information attached to these retweets, we identify the followee of each tweet. For each such followee, we also identify the followers who retweeted the message, and who did not. Finally we calculate the values of the corresponding features, and we get 621,261 retweet samples and 615,823 non-retweet samples after preprocessing such as duplicate removal and normalization (values are normalized into the range [0,1]). The ratio of the positive and negative samples is roughly 1:1, which we deliberately to do in order to handle class balance because of having enough data. We randomly split the sample set into training set (containing 80 % data) and testing set (containing 20 % data).

3.3 Predictive Models with All Features

Predicting is a binary classification task. Each sample is assigned a label 0 (will not be retweeted) or 1 (will be retweeted). We use three metrics to assess prediction accuracy: precision, recall and F-Measure (F1 score).

We use 51 features in total, including 15 followee features, 5 tweet features, 15 follower features and 16 interaction features. Through the calculation of eigenvalues, we get a 52 dimensional feature vector from each sample, and each component represents the corresponding eigenvalue and the last component is a label (0 or 1).

Table 1. The performance of classifiers with 51 features on negative & positive samples

Classifier	Precision(N)	Recall(N)	F1(N)	Precision(P)	Recall(P)	F1(P)
BayesNet	0.871	0.956	0.912	0.952	0.86	0.904
IBk	0.946	0.973	0.96	0.973	0.945	0.959
LR	0.781	0.828	0.804	0.818	0.77	0.793
RandomForest	0.967	0.986	0.976	0.986	0.967	0.976
J48	**0.976**	**0.995**	**0.986**	**0.995**	**0.976**	**0.986**

Table 1 shows different performance of five classifiers on the task of predicting who will retweet or not. These classifiers use all the features introduced above.

We find that the performance of BayesNet, IBk and Logistic on negative samples are better slightly than their performance on positive samples. The performance of RandomForest and J48 on negative samples are as the same as their performance on positive samples. It is worth noting that using J48 gives us the best performance.

3.4   Features Rank

Information gain [14] measures the number of bits of information obtained for category prediction by knowing the presence or absence of a term in a document. The information gain of term t and category c_i is defined to be:

$$IG(t, c_i) = \sum_{c \in \{c_i, \bar{c}_i\}} \sum_{t' \in \{t, \bar{t}\}} P(t', c) \cdot \log \frac{P(t', c)}{P(t') \cdot P(c)} \tag{2}$$

Using the above formula, we get the information gain value of each feature and rank these features in Table 2. By comparing followee features and follower features, we find that the top six features in this table are all followee features and followee features generally rank higher than follower features, which reveals the discrimination of followee features is stronger than follower features for predicting retweeting. The highest one in follower features is the seventh feature. The countOfStatuses-ratio, a interaction feature indicating the ratio of two numbers of tweets the followee and his/her follower posted, gets a high score. The count of words in a tweet is also a good feature for predicting retweeting.

3.5   Predictive Models Combining Top 12 Features

Based on the rank of characteristic importance in Table 2, we take the top 1 feature, the top 2 features, ..., the top 51 features in turn to train the model J48 and test F-Measure, and we use the results to draw Fig. 1.

When selecting the top several features, the experimental results should be poor theoretically. But why the test results show very good in Fig. 1? Because of selecting less features, duplicate sample data get more and we need to remove duplicate sample data further. But negative and positive samples data are seriously unbalanced this moment. Then we train and test the model J48 with these data, and the model will predict all the testing set as only one class, which making the results good but making little sense. When selecting the top 8 features, we get 78,551 negative samples and 348,905 positive samples. When selecting the top 9 features, we get 189,503 negative samples and 386,174 positive samples. The number of positive samples is probably double than that of negative samples, and its test result may be considered but not high. Along with the increasing of features selected, we find that its F-Measure is the highest combining the top 12 features and the results change little after that.

Therefore, we select the top 12 features containing 8 followee features, 2 follower features, 1 tweet feature and 1 interaction feature, and their information gain values are all above 0.2. We use the above five models to train and test

Table 2. The rank of 51 features

Rank	Value	Feature	Feature type
1	0.67632224834	countOfFollowers	Followee
2	0.67272397255	UI	Followee
3	0.66938223867	countOfStatuses	Followee
4	0.56668758481	countOfFriends	Followee
5	0.56553721581	days	Followee
6	0.5146334176	countOfBi-followers	Followee
7	0.3574061351	countOfStatuses*	Follower
8	0.29301535422	countOfStatuses-ratio	Interaction
9	0.29101604674	days*	Follower
10	0.27644664725	countOfFavourites	Followee
11	0.27014883961	lengthOfDescription	Followee
12	0.23186336321	lengthOfWeibo	Tweet
13	0.19083291706	countOfUrl	Tweet
14	0.17600458998	countOfFollowers*	Follower
15	0.1430282011	countOfFriends*	Follower
16	0.12931032668	lengthOfName	Followee
17	0.10582461682	countOfFollowers-ratio	Interaction
18	0.10351842687	days-ratio	Interaction
19	0.1024747354	followers-friends-ratio	Interaction
20	0.08748018856	countOfBi-followers-ratio	Interaction
21	0.08693831043	lengthOfName-ratio	Interaction
22	0.07943772309	countOfHashtag	Tweet
23	0.07413644717	countOfAt	Tweet
24	0.06750722606	countOfBi-followers*	Follower
25	0.06688156827	countOfFriends-ratio	Interaction
26	0.06068202909	lengthOfDescription-ratio	Interaction
27	0.04789192071	province	Followee
28	0.04786854032	UI*	Follower
29	0.0455341243	countOfFavourites*	Follower
30	0.04465197059	distance	Interaction
31	0.04429440811	verified-both	Interaction
32	0.03972143287	verified	Followee
33	0.03570689635	lengthOfDescription*	Follower
34	0.03479733741	lang-both	Interaction
35	0.03434400742	lang	Followee
36	0.03296798905	city	Followee
37	0.02548126363	countOfFavourites-ratio	Interaction
38	0.02525829954	countOfShareRel	Interaction
39	0.02163512059	url-both	Interaction
40	0.02121072136	url	Followee
41	0.02047033515	is-bi-follower	Interaction
42	0.01593060344	gender-both	Interaction
43	0.01171595753	gender*	Follower
44	0.00556299078	verified*	Follower
45	0.00304261666	countOfPic	Tweet
46	0.00252591821	province*	Follower
47	0.00239673417	lengthOfName*	Follower
48	0.00093883209	city*	Follower
49	0.00010915641	lang*	Follower
50	0.00001041463	url*	Follower
51	0.0000000028	gender	Followee

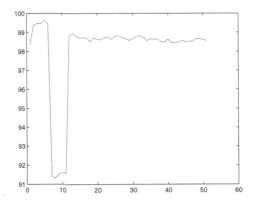

Fig. 1. The performance of the model J48 along with the increase of features one by one. Here vertical axis represents the F1 scores and horizontal axis represents the count of features from 1 to 51.

as well. Because we only use these 12 features, the sample set need to further process. We get 599,958 positive samples and 573,638 negative samples after removing duplicate data. Table 3 shows that J48 is still the best one of all and its result is better sightly than before, which may resulting from data reduction. So we draw the conclusion: J48 is the best predictive model with the top 12 features.

Table 3. The performance of classifiers with 12 features on negative & positive samples

Classifier	Precision(N)	Recall(N)	F1(N)	Precision(P)	Recall(P)	F1(P)
BayesNet	0.866	0.982	0.92	0.98	0.854	0.913
IBk	0.975	0.986	0.981	0.986	0.976	0.981
LR	0.791	0.772	0.782	0.786	0.804	0.795
RandomForest	0.977	0.995	0.986	0.995	0.977	0.986
J48	**0.978**	**0.999**	**0.988**	**0.999**	**0.978**	**0.988**

4   Conclusions

Retweeting is a key mechanism for information progagation in microblogs. To understand what features playing a key role in retweeting prediction, we investigated a total number of 51 features that have potential relationship with the retweetability. By comparing five classifiers with 51 features, we identified the best model. We also got 12 most important features by means of information gain method and the best classifier's experiments. For a given tweet of a given user, we could use our predictive model to predict who will retweet or not among his/her followers with up to 98 % accuracy.

Nevertheless, there are other aspects worth investigating in this work. First, because the contribution of each feature to the classification is different, we may consider to assign weights to different features. Second, we may further study tweets content, for example, what is the theme of a tweet, and their followers prefer to forward which type of tweets.

References

1. Zaman, T.R., Herbrich, R., Van Gael, J., Stern, D.: Predicting information spreading in Twitter. In: Workshop on, Computational Social Science and the Wisdom of Crowds (2010)
2. Krishnamurthy, B., Gill, P., Arlitt, M.: A few chirps about Twitter. In: 1st Workshop on Online Social Networks, pp. 19–24. ACM Press, Washington (2008)
3. Petrovic, S., Osborne, M., Lavrenko, V.: RT to win! predicting message propagation in Twitter. In: 5th International AAAI Conference on Weblogs and Social Media, California (2011)
4. Li, Y., Chen, Y.H., Liu, T., Deng, W.C.: Predicting the popularity of messages on micro-blog services. In: Huang, H., Liu, T., Zhang, H.-P., Tang, J. (eds.) SMP 2014, vol. 489, pp. 44–54. Springer Press, Beijing (2014)
5. Lee, K., Mahmud, J., Chen, J.: Who will retweet this?: automatically identifying and engaging strangers on Twitter to spread information. In: 19th International Conference on Intelligent User Interfaces, pp. 247–256. ACM Press, Israel (2014)
6. Webberley, W., Allen, S., Whitaker, R.: Retweeting: a study of message-forwarding in Twitter. In: Workshop on, Mobile and Online Social Networks (2011)
7. Suh, B., Hong, L., Pirolli, P., Chi, E.H.: Want to be retweeted? large scale analytics on factors impacting retweet in Twitter network. In: 2nd IEEE International Conference on Social Computing (socialcom), pp. 177–184. IEEE Press, Minneapolis (2010)
8. Yang, J., Counts, S.: Predicting the speed, scale, and range of information diffusion in Twitter. J. ICWSM **10**, 355–358 (2010)
9. Ota, Y., Maruyama, K., Terada, M.: Discovery of interesting users in Twitter by overlapping propagation paths of retweets. In: 2012 IEEE/WIC/ACM International Joint Conferences on Web Intelligence and Intelligent Agent Technology, pp. 274–279. IEEE Press, Macau (2012)
10. Cha, M., Haddadi, H., Benevenuto, F., Gummadi, P.K.: Measuring user influence in Twitter: the million follower fallacy. J. ICWSM **10**, 30 (2010)
11. Ho, C.T., Li, C.T., Lin, S.D.: Modeling and visualizing information propagation in a micro-blogging platform. In: Advances in Social Networks Analysis and Mining, pp. 328–335. IEEE Press, Washington (2011)
12. Bao, P., Shen, H.W., Huang, J., Cheng, X.Q.: Popularity prediction in microblogging network: a case study on Sina Weibo. In: 22nd WWW International Conference on World Wide Web Companion, pp. 177–178. ACM Press, Brazil (2013)
13. Hall, M., Frank, E., Holmes, G., Pfahringer, B., Reutemann, P., Witten, I.: The WEKA data mining software: an update. ACM SIGKDD Explor. Newsl. **11**(1), 10–18 (2009)
14. Zheng, Z.H., Wu, X.Y., Srihari, R.: Feature selection for text categorization on imbalanced data. ACM SIGKDD Explor. Newsl. **6**, 80–89 (2004). New York

Predicting User Relationship from Scratch

Xiaoxia Liu[✉], Hongfei Lin, and Zhihao Yang

School of Computer Science and Technology, Dalian University of Technology,
Dalian 116024, Liaoning, China
liuxiaoxia@mail.dlut.edu.cn, {hflin,yangzh}@dlut.edu.cn

Abstract. A task of primary importance for social network users is to differentiate whom and what to trust among large information. The trustworthiness of the users is often tantamount to the reliability of the information they provide. In this paper we focus on automatic methods for assessing the credibility of a given pair of users. Specifically, we establish a model to classify them as credible or not credible, based on features extracted from them. We test our model on three real world dataset Epinions, Slashdot and Wikipedia, the results indicate that although we only knew tiny information about a user, we can infer their relationship with higher accuracy compare to other researchers.

Keywords: Social networks · Trust · Em algorithm · User features

1 Introduction

With the rapid development of social networks, there are more and more people use internet to create content and share information. For example, More than 1 billion unique users visit YouTube each month and about 72 hours of video are uploaded to YouTube every minute. As so much user content created, it is necessary for users to discover, evaluate, and select sources of information that are worth their attention from a vast pool of potential choices. Trust, which provides information about from whom we should accept information, and with whom we should share information, can help a user to make decisions, sort and filter information.

Fortunately, the emergence of social networks has allowed users to indicate whom they trust and distrust, creating links between users in the network. For instance, In the Epinions platform users can flag another user as "trustworthy" or "untrustworthy". However, a given user likely knows only a small fraction of the people with whom he/she will interact in a large network, thus the user has no knowledge of how trustworthy most people are. To handle this, methods are needed for inferring trust between users who do not know one another directly.

There are a number of existing algorithms for inferring trust in social networks [1–4], roughly divided into two groups: unsupervised methods [1,3] and supervised methods [2,4]. Unsupervised methods such as trust propagation [1] and path- probability [3] can infer trust relations for two indirectly connected users. However, the

© Springer Science+Business Media Singapore 2015
X. Zhang et al. (Eds.): SMP 2015, CCIS 568, pp. 176–183, 2015.
DOI: 10.1007/978-981-10-0080-5_16

power law distribution indicates that the available trust relations may not make great effect to assure the success of these methods [4, 5]. While supervised method predict uses relationship by extracting features from available sources. Because of the lack of available information, the question of finding suitable features becomes difficult. In [2], they use features based on theories of social balance combine with user degree features, their results show that the signs of links between users can be predicted with high accuracy. We go a step further than existing works by exploring more efficient and effective user features, especially our model can predict negative relationship between users with a higher accuracy which is found more harder to improve when use real and unbalanced data.

In additionally, many computational approaches use the vast sociological and psychological literature on human behavior as their foundations. Recent studies have analyzed the structural balance- and status-based models in the social media data [2], and some researches also considered social theories of homophily effect for trust prediction [6]. In this paper we focus on the homophily theory which is one of the most important theories that attempt to explain why people establish trust relations with each other [7]. As the homophily effect suggests that similar users have a higher likelihood to establish trust relations, we use trust score which is computed by using trust score-rank algorithm to indicate users similarity globally. The trust score-rank algorithm adapts Geo-locating trust algorithm [8] in order to deal with the problem of correctly estimating the trustworthiness of the users, globally computes a ranking of the users which represents users trustworthiness in a social network.

We contribute to this area a new algorithm which regarding both users' global and local features for effectively predicting trust and distrust in social networks. We evaluate the performance of our algorithm on three real-world datasets, and the experimental results show accuracy and universality of the algorithm. The remainder of this paper is organized as follows: In Sect. 2, we review related works. Next, Sect. 3 introduces the dataset used in the experiments. We introduce our models as well as the estimation algorithm in Sect. 4. We present the experimental results in Sect. 5. Finally, conclusions and future work are discussed in Sect. 6.

2   Related Work

People form links, both positive and negative, to indicate friendship or hostility, support or disapproval in a social network. For a given link in a network, there are positive and negative edges according to whether the edge expresses a positive or negative attitude from the link generator to link recipient. Works dealing with the signed networks are [2,3,9,10]. In [10], they analyze the different between users negative and positive interaction, and which characteristics have effect on the networks with positive and negative implication, whereas in [2,3,11], authors studied the problem of classifying the edge sign in a social networks.

In a signed or unsigned network, user has his/her own trustworthiness in a global view. How to define a reasonable measurement to distinguish different user trustworthiness, as well as how to handle the negative trust score in a signed network and last but not least how to maintain users true trustworthiness against some adversarial or spam users. Recently, many PageRank and HITS variants, such as reputation-based ranking [8] and PolarityRank [8] algorithms, have been proposed to deal with negative edge problem. Furthermore, some algorithms also compute both prestige and bias of nodes in trust network [12]. To the best of our knowledge, none of them is used as a node characteristic as we do to predict trust relationship between users.

3   Data Description

In this section, we briefly introduce the data of our experiment. Following [2] we used three large online social network datasets to test our methods. All the networks where each link is labeled as positive (trust) and negative (distrust) are provided by the Sanford Large Network Datasets Collection1.

Epinions is a product website where users can write reviews. Users may author reviews, rate the reviews of other authors, and they are connected with trust or distrust based on the ratings and reviews. The network has 119,217 nodes and 841,200 edges, of which 85 % are positive. About 80,000 uses received at least one link edge, while over 49,000 users have both created and received a link from another user.

Slashdot is a technology news site with news updated every day. Users can comment the news as well as can rate each other as friend or foe. In this signed network as Epininons, friendship means that a user likes another user's comments, while a foe relation means the opposite way. We treat those as positive and negative trust ratings. The dataset contains over 82,144 users and 549,202 links of which 77.4 % are positive. Among them over 70,000 users received one trust or distrust edges and there are about 32,000 users with at least one in- and out- link.

Wikipedia provides a multitude of opportunities for large-scale online knowledge collaboration is the largest and most popular general reference work on the Internet. It has a set of elected moderators who help maintain it by monitoring the site for controversy and quality. The network we study corresponds to the election of Wikipedia users for promoting a user whether to admit the moderator. A signed link in this network indicates a positive or a negative vote by one user on the election of another user. There are 7,118 nodes and 107,080 edges of which 78.7 % are positive in this network and about 2,700 nodes received at least one edge and over 1,300 users created and received non-zero links.

4   Predicting User-Relation

For a given directed network $G < V, E >$, in which V is a set of nodes and E is a set of directed edges, with an edge sign either positive or negative, we use $s(u, v)$ indicates the sign of edge $e(u, v)$ from u to v. If the sign of $e(u, v)$ is positive then $s(u, v) = 1$, while if the sign of $e(u, v)$ is negative then $s(u, v) = -1$, otherwise $s(u, v) = 0$.

4.1 Node Features

Node-Degree Feature. Each user in a directed network has its own features, in which some of them are easily to be accumulated. We extracted two groups of users' local features based on their degree as follows to predict the sign of $s(u, v)$ between u and v. The features of first group as G1 which is like Leskovec [2] include *pout* (positive out-degree of a node), *nout* (negative out-degree of a node) and *tout* (total out-degree of a node) of node u, and *pin* (positive in-degree of a node), *nin* (negative in-degree of a node) and *tin* (total in-degree of a node) of v, together with $CM_{u,v}$ which denote the number of neighbors that u and v have in common in a undirected sense. The second group features as G2 consist of *noto* and *niti* of u and v as: $noto = nout/tout$ and $niti = nin/tin$. These two features present the negative features of a node which make sign predicting more accuracy when predict the negative relation between users.

Node-Topology Feature. As [2], we consider each triangle involving the edge $e(u, v)$, consisting of a node w with two edges one of them is either from u or to u and the other edge is also either from v or to v. Since the edge sign has two different values either positive or negative, so there will be 2*2*2*2=16 possible triangle type composed by three nodes with the edge in either direction and of either sign between u and w and also with the edge between w and v. Compare to balance theory and status theory, each of the trial type provides different evidence about the sign of edge of u and v, some tending a negative sign and some tending a positive sign. So we use 16 features as G3 based on node topology to form a 16-dimensional vector to describe the trial.

Node-Global Feature. Based on the social theory of homophily, we know that users with higher similarity have a higher likelihood to trust each other. We use confidence score to indicate users credibility globally which is calculated by the algorithm as shown in Table 1.

$$score_n = I_n + \prod_{i=0}^{n} D^{max-i} score_{max} \quad (0 < D < 1) \tag{1}$$

D denotes a decreasing parameter, which is related to *pin* of current node and the distance between current node and the node with highest *pin*. Then, we use G1 features and users' confidence score to compute the similarity between users and we use EM algorithm [13] to cluster uses into different groups, for example, user u belongs to $cg(u)$.

4.2 Method

Since the triad features are relevant only when u and v have neighbors in common, it is natural to expect that they will be most effective for edges of greater embeddedness ($CM_{u,v}$). We therefore consider the performance restricted to subsets of edges for different levels of minimum embeddedness. Based on minimum value of $CM_{u,v}$, each dataset has three conditions: embed0, embed10 and

Table 1. Algorithm of computing users' confidence score

Algorithm:	Computing users' confidence score
Input:	A directed network and initial confidence score I_n
Output:	Nodes' confidence *score* value
Step1:	Select the nodes with highest *pin* and assign I_n as their initial confidence *score*
Step2:	From the nodes with highest *score*, use breadth-first search algorithm (the direction of search algorithm is from positive) and Eq. 1 to calculate confidence score of other nodes, and for a node v if the new *score$_v$* is unequal to the old *score$_v$*, check the total *score* of v' neighbors and update *score$_v$*
Step3:	Check if each node has been searched, if so go to step 4; otherwise select nodes with highest *pin* and assign a new I_n as their confidence *score*, then go to step 2
Step4:	Output *score* of each node

embed25. We use a logistic regression classifier to combine users' local and global features with users' group into an edge sign prediction. Logistic regression learns a model of the form

$$P(s|\mathbf{x}) = \frac{1}{1 + e^{-(b + \sum_{i=1}^{n} b_i x_i)}} \tag{2}$$

where $\mathbf{x}$ is a vector of features $(x_1, \ldots, x_n)$ and $b_0, \ldots, b_n$ are the parameters we estimated based on the training data. The specific experimental steps are as follows: First, we utilized EM cluster algorithm using G1 and users' confidence *score* features to cluster users into different groups. Second, we used 12 feature classes for our machine-learning approach to this problem. For a unknown relation from user u to v, Table 2 shows the details of each feature vector used to predict the sign between them.

5   Experimental Results

We used leave-one-out cross-validation to predict the sign of edges. Moreover, accuracy and the area under the ROC curve (AUC) were used as evaluation metrics. The classification accuracy is shown in Fig. 1, where results are described for the three datasets, for the 12 classes features separately, and for different levels of minimum $CM_{u,v}$. Several observations stand out. First, the best accuracy of classification was by using G11 which combines both local and global features of users. Compared to the result of G7, which is the best of [2] for embed25, the accuracy increases 1.28 % (Epinions), 1.76 % (Slashdot) and 2.34 % separately. In addition, Table 3 shows the accuracy of G11 and G7 (Baseline) for embed25. This suggests that edge signs can be meaningfully in terms of both local and global properties, and these are without complex information such as gender or status of users. Second,the G0 feature performs well than the G1 and G3 features.

Table 2. Features of different feature vectors

Name	Including features
G0	$cg(u)$, $cg(v)$
G1	$CM_{u,v}$, $pout_u$, pin_u, $tout_u$, pin_v, nin_v, tin_v
G2	$noto_u$, $noto_v$, $nini_u$, $nini_v$
G3	16 different conditions from u to v through w
G4	G1+G2
G5	G0+G1
G6	G0+G1+G2
G7	G1+G3
G8	G2+G3
G9	G0+G1+G3
G10	G1+G2+G3
G11	G0+G1+G2+G3

This indicates that the cluster of users can play an essential part in predicting users' relation. Finally, it is also noteworthy with G2 the accuracy increases significantly when compare the results of G8 and G10 with G9 and G11. This shows *noto* and *niti* are useful for edges prediction, especially negative sign prediction.

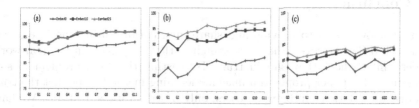

Fig. 1. Accuracy of predicting trust between users by using different feature vectors (a,b,c corresponds to Epinions, Slashdot, and Wikipedia respectively).

In all experiments we reported the average accuracy and estimated logistic regression coefficients over 10-fold cross validation. If not stated, we limit our results with minimum embed25. When evaluating using AUC rather accuracy, the various forms of logistic regression have AUC of approximately 97 % on the three datesets. To estimate the impact of cluster number on edge prediction, we analyze the accuracy of G5 when use different cluster number, as illustrated on Fig. 2. We can note that the accuracy gradually increases with the increase of cluster number. However, after reaching a certain value it presents downtrend. This may be due to the increasing number of cluster which will lead to unobvious

Table 3. Accuracy of predicting trust and distrust on three datasets

Accuracy	Epinions	Slashdot	Wikipedia
Trust accuracy of G11	98.27 %	94.87 %	94.93 %
Distrust accuracy of G11	62.68 %	55.53 %	50.53 %
Total accuracy of G11	**93.04 %**	**85.97 %**	**85.35 %**
Total accuracy of G7	91.36 %	83.88 %	81.28 %

differentiation between users. We roughly choose twenty as the clustering number of all datasets.

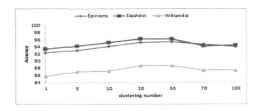

Fig. 2. The relationship between clustering number and trust prediction accuracy

6   Conclusion

Internet allows users to provide and share information during the interaction with each other, however, they lack the clues that they have in the real world to determine whom they should trust. By casting this as a problem of sign prediction, this paper proposed a machine learning method to predict the trust and distrust relationship between users, only based on users' features. We have shown that similar users have a higher likelihood to establish trust relations. According to these feature and other local prosperities of users, we can predict user relation automatically with high accuracy. We also test the robust of our model, it suggests the accuracy of prediction doesn't depend on the training dataset.

There are a number of further implementation can be done in future. For example, this work can be used to filter junk message from distrust users, and to ensure safety of internet, and to recommend things to people liked by they trust. For future work, we plan to extend the experiments to undirected network, to balance network (eg. with equal number of positive and negative edges), and to explore more effective features.

Acknowledgments. This work is partially supported by grant from the Natural Science Foundation of China (No. 61277370, 61572102), Natural Science Foundation of

Liaoning Province, China (No. 201202031, 2014020003), State Education Ministry and The Research Fund for the Doctoral Program of Higher Education (No. 20090041110002).

References

1. Guha, R., Kumar, R., Raghavan, P., Tomkins, A.: Propagation of trust and distrust. In: Proceedings of the 13th International Conference on World Wide Web, (ACM 2004), pp. 403–412 (2004)
2. Leskovec, J., Huttenlocher, D., Kleinberg, J.: Predicting positive and negative links in online social networks. In: Proceedings of the 19th International Conference on World Wide Web, (ACM 2010), pp. 641–650 (2010)
3. DuBois, T., Golbeck, J., Srinivasan, A.: Predicting trust and distrust in social networks. In: 2011 IEEE Third International Conference on Privacy, Security, Risk and Trust (PASSAT) and 2011 IEEE Third Inernational Conference on Social Computing (SocialCom), (IEEE 2011), pp. 418–424 (2011)
4. Tang, J., Chang, S., Aggarwal, C., Liu, H.: Negative link prediction in social media. In: Proceedings of the Eighth ACM International Conference on Web Search and Data Mining, (ACM 2015), pp. 87–96 (2015)
5. Tang, J., Gao, H., Liu, H., Das Sarma, A.: etrust: Understanding trust evolution in an online world. In: Proceedings of the 18th ACM SIGKDD International Conference on Knowledge Discovery and Data Mining, (ACM 2012), pp. 253–261 (2012)
6. Tang, J., Gao, H., Hu, X., Liu, H.: Exploiting homophily effect for trust prediction. In: Proceedings of the Sixth ACM International Conference on Web Search and Data Mining, (ACM 2013), pp. 53–62 (2013)
7. Liu, H., Lim, E.-P., Lauw, H.W., Le, M.-T., Sun, A., Srivastava, J., Kim, Y.: Predicting trusts among users of online communities: an epinions case study. In: Proceedings of the 9th ACM Conference on Electronic Commerce, (ACM 2008), pp. 310–319 (2008)
8. Ortega, F.J., Troyano, J.A., Cruz, F.L., Vallejo, C.G., EnríQuez, F.: Propagation of trust and distrust for the detection of trolls in a social network. Comput. Netw. **56**(12), 2884–2895 (2012)
9. Seckler, M., Heinz, S., Forde, S., Tuch, A.N., Opwis, K.: Trust and distrust on the web: user experiences and website characteristics. Comput. Hum. Behav. **45**, 39–50 (2015)
10. Szell, M., Lambiotte, R., Thurner, S.: Multirelational organization of large-scale social networks in an online world. Proc. Nat. Acad. Sci. **107**(31), 13,636–13,641 (2010)
11. O'Doherty, D., Jouili, S., Van Roy, P.: Towards trust inference from bipartite social networks. In: Proceedings of the 2nd ACM SIGMOD Workshop on Databases and Social Networks, (ACM 2012), pp. 13–18 (2012)
12. Li, R.-H., Yu, J.X., Huang, X., Cheng, H.: A framework of algorithms: computing the bias and prestige of nodes in trust networks. PLoS One **7**(12), (2012)
13. Moon, T.K.: The expectation-maximization algorithm. IEEE Sig. Process. Mag. **13**(6), 47–60 (1996)

Detecting Overlapping and Hierarchical Communities in Complex Network Based on Maximal Cliques

Zhenhua Huang, Zhenyu Wang[(⊠)], and Zhiwei Zhang

School of Software Engineering, South China University of Technology, Guangzhou, People's Republic of China
zhhuangscut@gmail.com, wangzy@scut.edu.cn

Abstract. Community detection is a fundamental task in discovering complex network. Maximal cliques are found to play significant roles in communities. This paper proposes an efficient algorithm OMC for detecting communities based on maximal cliques. Subordinate maximal cliques are removed and remaining cliques are regarded as initial communities. Then small communities are merged into larger ones according to a fitness function. The proposed algorithm is able to uncover both overlapping and hierarchical communities in high speed. Experimental results on various real-word networks have proven that the method performed well in detecting communities.

Keywords: Community detection · Maximal cliques · Overlapping · Hierarchical · Fitness function

1 Introduction

Real-word systems are often formed in terms of complex networks or graphs. Examples are biological networks, co-authorship network, protein interaction graphs, social network [1, 2]. Nodes in networks represent entities and links represent relations between entities. Community is found universal existing in complex network. A community might correspond to a certain functional unit in a protein network or a group of students study together in social network. However, there is not a uniform definition of community. Intuitively, community is the structure in complex network that has density links within but less connections to other communities.

A great deal of methods or algorithms have been proposed to detect communities in recent years, making it a hot interdisciplinary problem. Traditional algorithms provide a partition of network without overlapping. However, the overlapping information is later found significant in community detection. A node can play various roles in networks and belongs to several communities. For example in social network, a node represents a teacher might be head teacher in a class and also a member of faculty staffs. Palla proposed the first method CPM in which could detect overlapping communities. After Palla's work, maximal cliques have been widespread concern in community discovery.

Significant roles that cliques play in communities have been pointed out in details in this paper. Density parts of complex network tend to form larger cliques.

© Springer Science+Business Media Singapore 2015
X. Zhang et al. (Eds.): SMP 2015, CCIS 568, pp. 184–191, 2015.
DOI: 10.1007/978-981-10-0080-5_17

Basic structures of many communities are composed of maximal cliques. Subordinate cliques are proposed to remove disturbed maximal cliques. The paper proposes an algorithm based on Optimization over Maximal Cliques (OMC), which could detect overlapping community and also give hierarchical structure of communities. The algorithm applies the object function of fitness and evaluation function of extended modularity. Most methods based on optimization usually demand high time complexity. The algorithm in this paper keeps a good shape of community while greatly reducing time complexity. Experimental results indicated that the proposed algorithm OMC performed well in both the small-scale and large-scale data sets in real world.

2 Related Works

Community detection algorithms can be classified into node-based and link-based algorithms when terms to aspect of community member. Link-based algorithms are not focus in this paper. The majority of community detection algorithms are node-based. Node-based algorithms include overlapping and non-overlapping algorithms according to whether they allow overlapping nodes or not. The first traditional no-overlapping algorithms is GN. According to natural property of overlapping in cliques, Palla [2] presented first overlapping community detection algorithm CPM. However CPM will result in many small scattered cliques or very large cliques. GCE [3] applied part cliques as seeds and expanded them according to the fitness function to get community structure. This method had been proven performed well in high-density network. EAGLE [4] defined the similarity of communities and combined a pair of communities at each step. Coupling strength was applied by MOHCC [5] to assess the tightness of two communities. EAGLE and MOHCC produced a dendrogram and cut through it when evaluation function got maximum value. ACC [6] employed local function of clustering efficient to merge two neighboring complete sub-graphs. Lancichinetti [7] came up with first algorithm LFM which could detect overlapping and hierarchical communities. Recently, Coscia [8] proposed a local-first approach to uncover hierarchical and overlapping communities, with a limited time complexity, which can be used on large-scale networks.

3 Motivation

The skeleton of a community are constituted by several maximal cliques. Some maximal cliques might contain core members of communities. However, a simple method to extract maximal cliques is not enough to get good community structure, and a suitable method is required to combine cliques together, otherwise fragmented or too large communities will be obtained like CPM algorithm. The fitness function used in the paper leads density-linked cliques to merge with higher priority. It can produce communities with less links outside but with more links inside. Merging two cliques at one step led to high time complexity. Local fitness function is applied to improve the efficiency of merging. Evaluation function is also required to judge when algorithm produces best division, otherwise all maximal cliques will be merged together.

3.1 Maximal Cliques

Clique, or complete graph, is an important structure in complex network. Nodes in density part of community tend to connect each other so as to form complete graphs K-clique means a clique with k nodes. Maximal clique is the clique that is not sub-graph of another clique. One of the famous algorithms to find out maximal cliques in a graph was proposed by Bron [9]. Some scholars modified and improved the algorithm after Bron. It should be noticed that nodes in some cliques already exist in other lager cliques. Those cliques can affect the effect of merging and mislead the algorithm to combine two density communities together, making it unable to reflect good community structure. The cliques are called 'subordinate cliques' in this paper. The subordinate cliques will be considered after main community structures are extracted, for they may contain overlapping information. To save filtering time, only cliques with k <= 4 will be removed. If the difference of in-degree and out-degree of a node is less than 1, the node is regarded as overlapping nodes between two communities. Especially, 2-cliques are abandoned by some algorithms. However, we hold the view that some of them contain the same node link two communities and might be overlapping nodes. Some points of 2-cliques only connect with very less members. For example some nodes are just members of a community, but would not like to interact with others. But they should not be ignored since the points might be the target points and should be kept. For example, teacher may find out students not active to other students.

3.2 Extended Modularity

Modularity was proposed by Newman [10] and was used to evaluating performance of community detection algorithm at first. Later it was used as a way to detect communities. Modularity is measured by following function:

$$Q = \sum_{1}^{k} (\sum_{i,j \in c} A_{ij} - d_i d_j / 2M).$$ (1)

Modularity is proposed before overlapping community algorithm. Suppose there are two k-cliques A and B and they share k-1 nodes. Modularity before merging is $Q_1 = 2(k - \sum_{1}^{k} k_i k_j / 2M)$ and after merging is $Q_2 = Q_1 - \sum k_s k_t / 2M$. Wherein s and t are nodes not connected in the new graph. It can be seen clearly that the total modularity is cut off. EQ is measured as following, wherein O_i is the number of communities a node belongs to.

$$EQ = \sum_{1}^{k} (\sum_{i,j \in c} A_{ij} - d_i d_j / 2M) / O_i O_j.$$ (2)

3.3 Merge

When maximal cliques are extracted, there will be too many scattered small groups. The cliques are regarded as initial communities should be expanded or merged together to constitute larger communities. A clique will combine with another one when local fitness is most increased. There are various fitness functions. The following fitness function is proven efficient in this paper, α is set to 1 in our algorithm. Only local values are required to calculate the variation of fitness using the following function. The following fitness function tends to produce communities with less edge outside.

$$F_s = k_{in}^s / (k_{in}^s + k_{out}^s)^{\alpha}. \tag{3}$$

3.4 Algorithm Description

The algorithm OMC is described in the following. It stops when EQ gets max in this paper. Readers can also choose other termination conditions. For example, the number of community reaches a certain value or tightness between the two communities less than a threshold value. So the algorithm can be modified to satisfy different needs. The basic steps of our algorithm are as following:

- Initially, all maximal cliques in a network are extracted, each are given a unique id.
- Remove out subordinate cliques. Reminders are regarded as the initial communities.
- For each community, changes of local fitness are calculated if it combines with its neighboring communities.
- Choose the max value to improve fitness function and merge communities. Make sure the value is positive or large than a threshold (here we set the threshold to 0.0). EQ is calculated at the same time.
- Continue the algorithm until there are no communities could be merged, which a dendrogram is obtained.
- The dendrogram are cut through to get best division when total EQ at maximum value.
- Overlapping nodes are calculated.

Time complexity is analyzed. Extracting maximal cliques in a network is an N-P hard problem. It is hard to assess its time complexity precisely. However, it seldom takes long time in a real-word network during to their sparsity. Even for email network with 1133 nodes, the processing time is less than a second. Filtering process demands $O(n)$ operations. Suppose there are m maximal cliques, which is usually several times of n. Each round a community needs k operations to calculate local changes of fitness and to be merged with neighbors. So time in a round is up to $O(km)$. Time will sharp down greatly after several turns. The algorithm will stop for several turns t even for a large network. Without considering extraction of cliques, the total time complexity will be $O(km)$. Compared with EAGLE and MOHCC our algorithm is much time-saving.

4   Experiments and Results

4.1   Bottlenose Dolphins Network

Some bottlenose dolphins in New Zealand were living together at fist. Later because of unknown reasons, these dolphins are divided into two groups group, one large and one small. The network contains 62 nodes and 159 edges network. Results of the algorithm OMC are as follows. EQ value reaches max value of 0.238 in the second round. At this time the network is divided into two communities, one large and one small, which are consistent with real network (Fig. 1).

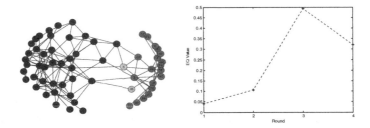

Fig. 1. Results on the bottlenose dolphins network. The left part shows communities, and the right part shows changes of EQ

4.2   Karate Club Network

Zachary's Karate Club Network is also one of widely used benchmarks for community detection. It contains 78 edges and 34 nodes. Disagreement of the coach and director led to network divided into two communities with center of coach and director respectively. Results are shown in Fig. 2. In the division, members tagged 3 and 10 are overlapping nodes and they are close to both communities, in line with the original findings of Zachary.

Results extracted by some other famous algorithm are shown in Fig. 3.

It can be seen that there are no overlapping nodes between communities in results of GN. And computational complexity of the GN algorithm is very high, making it unable to apply in large-scale networks. Some of communities extracted by CPM are not shown in the figure. The community showed in the Fig. 3 covers most nodes in the network, resulting in weakened sense of community division. GCE algorithm roughly divided the network into two communities but ignoring some nodes. Compared with these algorithms, our algorithm OMC can divide the club into two closely connected communities. Results of OMC here are similar to those of EAGLE and MOHCC.

Besides, hierarchical communities can be uncovered in karate club. In the first round of OMC, clique constitute a small community in the club, members '1', '11', '17', '5', '6', '7' constitute a small community in the club, members '1', '14', '2', '3', '4', '8' form another one. At the same time, the two small communities are connected by member '1' and they can form a larger community in the second round, which reveals

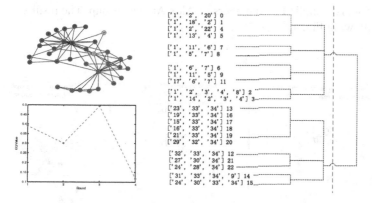

Fig. 2. Results on the karate club network.

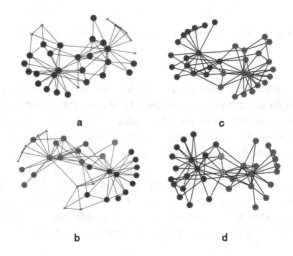

Fig. 3. Results extracted by some other algorithms on the karate club network. (a) CPM (b) GN (c) GCE (d) EAGLE

phenomenon of hierarchy in community. Whether in bottleneck dolphin network or karate club, maximal cliques play important roles in communities. In karate club, cliques tagged 14 and 15 maintain the basic framework of a community, while cliques tagged 2 and clique 3 forms framework of another one. In dolphins network, cliques ['19', '25', '30', '46', '52'] and ['15', '17', '34', '38'] contain most core members of the large community, and another clique ['10', '14', '18', '58', '7'] includes kernel members of the small community.

4.3   Email Network

The Email Network [11] contains Email interchanges between members of the Univerisity Rovira i Virgili. There are 1133 nodes and 10902 edges in the network.

The data set is compiled by members of Alex Arenas' group. The results are shown in the following Fig. 4.

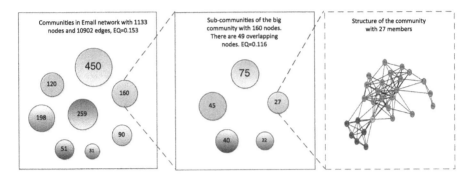

Fig. 4. Results on the Email Network (Color figure online).

The above results are obtained when OMC just runs for five rounds. Email Network is divided into eight big communities when EQ reaches the maximum value. Further dividing a community that contains 160 nodes, five smaller communities will be extracted. Structure of the sub-community with 27 nodes is shown in the figure. The yellow node represents overlapping node between communities that colored red and blue separately. It indicates that the algorithm OMC performs well in detecting community structures. Hierarchical information of communities can help us find composition relationships between communities. Overlapping information is helpful in analyzing which nodes act as contact people or middlemen between two or more communities. Both of them can be captured by proposed algorithm OMC.

5   Conclusion

Maximal cliques are highlighted to pay significance role in community detection in this paper. A novel algorithm OMC for community detection is proposed in this paper based on maximal cliques. In this algorithm, maximal cliques are first extracted as the initial communities. OMC can detect the high-density part in complex networks in lower complexity compared with other method based on maximal cliques. Communities extracted by OMC preserve complete information of original complex network and reveal the density linked structure exist in complex network. The algorithm not only uncovers the overlapping information, but also shows hierarchical structures in communities. The algorithm was proven high efficient and useful in different data sets.

References

1. Girvan, M., Newman, M.E.J.: Community structure in social and biological networks. Proc. Natl. Acad. Sci. U.S.A. **99**, 8271–8276 (2001)
2. Palla, G., Derenyi, I., Farkas, I., Vicsek, T.: Uncovering the overlapping community structure of complex networks in nature and society. Nature **435**, 814 (2005)
3. Lee, C., McDaid, A., Reid, F., et al.: Detecting highly overlapping community structure by greedy clique expansion. Paper presented at the 4th SNA-KDD Workshop 2010 (SNA-KDD 2010), 25 July, 2010, Washington, DC, USA (2010)
4. Shen, H., Cheng, X., Cai, K., Hu, M.: Detect overlapping and hierarchical community structure in networks. Phys. A Stat. Mech. Appl. **388**, 1706 (2009)
5. Zhang, Z., Wang, Z.: Mining overlapping and hierarchical communities in complex networks. Phys. A: Stat. Mech. Appl. **421**, 25 (2015)
6. Cui, Y., Wang, X., Li, J.: Detecting overlapping communities in networks using the maximal sub-graph and the clustering coefficient. Phys. A: Stat. Mech. Appl. **405**, 85 (2014)
7. Lancichinetti, A., Fortunato, S., Kertesz, J.: Detecting the overlapping and hierarchical community structure in complex networks. New J. Phys. **11**, 20 (2009)
8. Coscia, M., Rossetti, G., Giannotti, F., Pedreschi, D.: Uncovering Hierarchical and Overlapping Communities with a Local-First Approach. ACM Trans. Knowl. Disc. Data. **9**, 1 (2014)
9. Bron, C., Kerbosch, J.: Algorithm 457: finding all cliques of an undirected graph. Commun. ACM **16**(9), 575–577 (1973)
10. Newman, M.E., Girvan, M.: Finding and evaluating community structure in networks. Phys. Rev. E: Stat. Nonlin. Soft. Matter. Phys. **69**, 26113 (2004)
11. Guimera, R., Danon, L., Diaz-Guilera, A., Giralt, F., Arenas, A.: Self-similar community structure in a network of human interactions. Phys. Rev. E: Stat. Nonlin. Soft. Matter. Phys. **68**, 65103 (2003)

The Expert Ranking Method Based on Listwise with Associated Features

Sichao Wei[1,2,3], Zhengtao Yu[1,2(✉)], Fangqiong Chen[1], Cunli Mao[1,2], and Jianyi Guo[1,2]

[1] School of Information Engineering and Automation, Kunming University of Science
and Technology, Kunming, China
ztyu@hotmail.com
[2] Intelligent Information Processing Key Laboratory, Kunming University of Science
and Technology, Kunming, China
[3] Department of Industrial Information Technology, Yunan Vocational College of Mechanical
and Electrical Technology, Kunming, China

Abstract. In the expert ranking process, we put forward an expert ranking method based on list with associated features, in order to use effectively expert relationship and the relative relationship of the expert list. First, construct a correlation model of query and expert by evidence documents, expert display and implicit relationships and expert metadata. Then sort experts by the ListNet algorithm based on the list. Finally perform an experiment under the expert list, and the MAP value is 0.3506. The result shows that this method can effectively improve the accuracy of expert sorting, and expert relationship and the relative relationship of the expert list play an important role in the expert ranking.

Keywords: Expert ranking · Expert evidence document · Expert networks · Expert metadata · Expert-ListNet

1 Introduction

Expert retrieval is a hot area of the current vertical information retrieval. It has important application in every industry. The expert ranking model is the core of expert retrieval, and the merits of expert ranking determine the accuracy of expert retrieval. A lot of work has been carried out on the expert ranking both at home and abroad. Including, the ranking method based on expert file, for example, Macdonald et al. proposed using the frequency of name in the data set, the personal web page and the email to generate the description of the expert [1]; Merging method based on related documentation set, for instance, Fu et al. proposed a new method of document restructuring to collect and merger related information from different media formats, and generate a document for each expert [2]; Sort method based on document set. It is considered that the document is a bridge in expert

Supported by the China National Nature Science Foundation (No. 61175068, 61472168, 61163004), and The Key Project of Yunnan Nature Science Foundation (No. 2013FA130).

© Springer Science+Business Media Singapore 2015
X. Zhang et al. (Eds.): SMP 2015, CCIS 568, pp. 192–199, 2015.
DOI: 10.1007/978-981-10-0080-5_18

retrieval. Such as Balog et al. first sorted the document set according to the query subject, then found out the candidate list of experts through analysing the document collection of high relevance [3]. The expert probability model is presented by Fang, and it sorts the candidate experts based on the co-occurrence probability of themes and candidates in the supporting documents [4]. The above methods are based on the sort of data points. Actually, the sort of prediction experts is not based on a single expert, but the set of experts in the experts retrieval. The expert ranking method based on expert list has not been studied.

Most of the algorithms have little consideration of the relationship among the experts. Li et al. proposed expert ranking based on PageRank algorithm in enterprise micro blog [5]. For another example, Chen et al. used co-occurrence relationship among experts to establish expert relationship network, then adjust the scores of experts in accordance with the co-occurrence of experts networks [6]. This method has been integrated into the relationship among experts, and has achieved better results. It just considered simply the relationship among the experts, actually, the relationship among the experts is quite complex. For example, in the expert evidence document, expert display and implicit relational network and expert metadata, there are many relationship. These relationships have an important effect on the ranking of experts. So this paper studies how to make use of the related features to sort the experts, and discusses the expert ranking method based on list with associated features.

2   The Expert Ranking Method Based on List with Associated Features

2.1   The Correlation Model Based on Evidence Document

The relevance of query and expert can be reflected by the calculation of the relevance between query and document $p(r_1 = 1|q,d_t)$, the correlation between experts and documents $p(r_2 = 1|e,d_t)$, the prior probability $p(d_t)$ of document d_t, The model is as follows:

$$p_d(r = 1|e,q) = \sum_{t}^{n} p(r_1 = 1|q,d_t)p(r_2 = 1|e,d_t)p(d_t)$$

The prior probability $p(d_t)$ is used as constant. Calculating the correlation between query and document $p(r_1 = 1|q,d_t)$ uses BM25 algorithm with better effect in text retrieval. The correlation score of document d and query q is as follows:

$$score(d, q) = \sum_{t \in q} w \frac{(k_1 + 1)tfn}{k_1 + 1 + tfn} \frac{(k_2 + 1)qtf}{k_2 + qtf} \tag{1}$$

Because experts are characterized by the evidence documentation, the correlation between expert and document can be calculated document similarity by constructing the vector space model. In the vector space model, the content of document expressed by features. The similarity between the two texts can be represented by the cosine of the angle of the two text vector. Given document d_i and d_j, the similarity is defined as follows:

$$sim(d_i, d_j) = \cos(d_i, d_j) = \frac{\sum\limits_{k}^{n} w_{ik} \cdot w_{jk}}{\sqrt{\sum\limits_{k=1}^{n} w_{ki}^2} \sqrt{\sum\limits_{k=1}^{n} w_{jk}^2}} \cdot \qquad (2)$$

2.2 The Correlation Model Based on Expert Relationship Network

There are a variety of relationships among experts. Such as show relationships which contain partnership, mail relationship and so on, and implicit subject relationships which contain same domain relationship, research direction relation. These expert's relationships have the greater impact on the sort. So it can effectively use expert networks which has been constructed to calculate the correlation between expert and query. Then adjusts the sort results, and constructs the related model of query and expert based on expert relation net. It is as follows:

$$p_{net}(r = 1|e,q) = \sum_{i}^{m} p(r_1 = 1|q, NET_i)p(r_2 = 1|e, NET_i)p(NET_i)$$

m is the number of topics. NET_i represents the expert theme network which theme is i. $p(NET_i)$ indicates the connection strength of the network itself, and does not affect the ranking of the expert model, and can be recorded as a constant.

Calculating the correlation $p(r_1 = 1|q,NET_i)$ between query q and subject relation network is equivalent to find the relation of query q and keyword w. It can be calculated by HowNet [7]. For query terms q and subject terms w, If q there are n senses: S_{11}, S_{12}, ...,S_{1n}, w there are m senses: $S_{21}, S_{22}, ...,S_{2m}$, We provision the relationship between q and w which is the maximum similarity of the concepts. That is to say:

$$p(q, w) = \max_{i=1...n,j=1...m} Sim(S_{1i}, S_{2j}) \qquad (3)$$

The correlation $p(r_2 = 1|e,NET_i)$ between expert e and subject relation network can be get by the calculation of the co-occurrence relations between expert and subject words [8].

2.3 The Correlation Model Based on Expert Metadata

Expert metadata is structured attribute data and has a lot of accurate description of expert. Therefore, the retrieval ranking can be carried out by using the correlation of the query, expert and these structural data. So we extracted the basic metadata of experts to construct expert metadata repository in the process of constructing expert resource. Its ranking model is as follows:

$$p_{meta}(r = 1|e,q) = p(r_1 = 1|q,A)p(r_2 = 1|e,A)p(A)$$

$p(r = 1|\bullet, A)$ is the relevance of and property A, P(A) is the prior probability of property A used as a constant.

The expert metadata repository includes four kinds metadata which are the name, unit, research direction and positional titles of expert, each type of metadata as a one-dimensional feature. Then define five levels which are particularly strong correlation, strong correlation, correlation, weak correlation and irrelevance to represent the correlation between query and expert metadata. The corresponding probability values are 1, 0.75, 0.5, 0.25 and 0. It follows that the correlation between query and the metadata has been transformed into a classification problem. First, we constructed the feature vector of the query in the training sample according to four types of metadata feature and marked one of five levels. Then, five kinds of classifiers are generated by the support vector machine SVM. The correlation between expert and expert metadata can be calculated by simple and effective linear weighted normalization method.

2.4   The Expert-ListNet Algorithm

On the basis of the above three models, use exponential to smooth the sorting discriminant model. It is as follows:

$$p(r = 1\,|e, q) = \exp\left(-\frac{1}{\alpha \cdot p_d(r=1|e,q)+\beta \cdot p_{net}(r=1|e,q)+(1-\alpha-\beta)\cdot p_{meta}(r=1|e,q)}\right) \quad (4)$$

Calculate the correlation between query and a single expert by the formula (4), and sort the experts according to the relevant degree. But this approach ignores the correlation between the expert list. Therefore, we use the way to sort expert based on recalling the expert list by the introduction of permutation probability [9]. That is, for a sample group of experts $E = \{e_1, e_2, \ldots, e_n\}$, the score of each expert in E was given by the discriminant model of expert, and it is $s_i = p(r = 1|e,q)$, then forms $S = \{s_1, s_2, \ldots, s_n\}$. π is defined as the arrangement of the expert list. $p_s(\pi)$ is the probability of occurrence of π by S. The calculation formula of $p_s(\pi)$ is as follows:

$$P_s(\pi) = \prod_{j=1}^{n} \frac{\Phi(S_{\pi(j)})}{\sum_{k=j}^{n} \Phi(S_{\pi(k)})} \quad (5)$$

Φ is a linear function, $\phi(x) = \alpha x, \alpha > 0$. $f_{\alpha,\beta}$ is ranking function, $f_{\alpha,\beta} = p(r = 1\,|e, q)$. Given a query $q^{(i)}, e_j^{(i)}$ is j-th expert in the specialist collections related to the query. Then feature vector is $x_j^{(i)} = \Psi(q^{(i)}, e_j^{(i)})$, and scoring list is $z^{(i)} = (f_{\alpha,\beta}(x_1^{(i)}), \ldots, f_{\alpha,\beta}(x_{n(i)}^{(i)}))$. By formulas (4) and (5) the order probability is:

$$P_{z^{(i)}(f_{\alpha,\beta})}(\pi) = \prod_{j=1}^{n} \frac{f_{\alpha,\beta}(x_j^{(i)})}{\sum_{k=j}^{n} f_{\alpha,\beta}(x_k^{(i)})} = \prod_{j=1}^{n} \frac{p(r = 1\,\big|e_j, q)}{\sum_{k=j}^{n} p(r = 1\,\big|e_k, q)} \quad (6)$$

For query $q^{(i)}$, using cross entropy to calculate the minimum loss function is as follows:

$$L(y^{(i)}, z^{(i)}(f_{\alpha,\beta})) = -\sum_{\forall g \in \pi} P_{y^{(i)}}(g) \log(P_{z^{(i)}(f_{\alpha,\beta})}(g))$$

$y^{(i)}$ is the correlation between query and the experts which is artificial marking. The gradient of minimum loss function with parameters α, β is as follows:

$$\Delta\alpha = \frac{\partial L(y^{(i)}, z^{(i)}(f_{\alpha,\beta}))}{\partial \alpha} = -\sum_{\forall g \in \pi} \frac{\partial P_{z^{(i)}(f_{\alpha,\beta})}(g)}{\partial \alpha} \frac{P_{y^{(i)}}(g)}{P_{z^{(i)}(f_{\alpha,\beta})}(g)} \qquad (7)$$

$$\Delta\beta = \frac{\partial L(y^{(i)}, z^{(i)}(f_{\alpha,\beta}))}{\partial \beta} = -\sum_{\forall g \in \pi} \frac{\partial P_{z^{(i)}(f_{\alpha,\beta})}(g)}{\partial \beta} \frac{P_{y^{(i)}}(g)}{P_{z^{(i)}(f_{\alpha,\beta})}(g)} \qquad (8)$$

Minimizes the loss function $L(y^{(i)}, z^{(i)})$ in the gradient descent algorithm using formula (7) and (8), then gets the best model parameters α and β. Finally, using ranking model trained sorts the experts. Algorithm 1 introduces the Expert-ListNet algorithm. This method calculates the uniform loss for all the experts under a given query, instead of creating a separate loss for each expert. Constructed expert retrieval ranking model, which integrates the related features of the correlation model based on evidence document, the correlation model based on expert relationship network, the correlation model based on expert metadata, ranks experts in the list. And it improves the performance of expert retrieval.

Algorithm 1. Learning Algorithm of Expert-ListNet

Input: training data {(x(1), y(1)), (x(2), y(2)), ..., (x(m), y(m))}

Parameter: number of iterations T and learning rate τ

Initialize parameter α, β

for t = 1 **to** T do

for i = 1 **to** m do

Input $x^{(i)}$ of query $q^{(i)}$ to model and compute score list $z^{(i)}(f_{\alpha,\beta})$ with current α, β

Compute gradient $\Delta\alpha$ and $\Delta\beta$ using Eq. (7) and (8)

Update $\alpha = \alpha - \tau \times \Delta\alpha$, $\beta = \beta - \tau \times \Delta\beta$

end for

end for

Output parameter α, β

---

3   The Experiment and Result Analysis

3.1   The Experimental Data Preparation

Since there is no open authority corpus resources in the expert retrieval, we collect 10,130 evidence documentations of experts by artificial collection. Documentations which include experts homepage, experts blog page, baidu encyclopedia, wikipedia pages and non-expert page, construct the document collection of the candidate and the query of expert retrieve [10]. We divide the data set into 5 parts, respectively as S1, S2, S3, S4 and S5, then process 5-fold cross validation. We select three parts of data set as the training set, one parts of data set as the validation set, one parts of data set as the test set in each fold, as shown in Table 1. In addition, the expert relationship network and expert metadata repository have been used in the experiment.

Table 1. Data segmentation of 5-fold cross validation

Folds	Training set	Validation set	Test set
Fold1	{S1, S2, S3}	S4	S5
Fold2	{S2, S3, S4}	S5	S1
Fold3	{S3, S4, S5}	S1	S2
Fold4	{S4, S5, S1}	S2	S3
Fold5	{S5, S1, S2}	S3	S4

3.2 The Influence of Different Correlation Characteristics to Expert Sort

In order to study the influence of the correlation characteristics in expert evidence documents, expert display and implicit relationships network and expert metadata, four groups of comparative experiments are set up, and they are as follows:

A: The correlation model based on evidence document;
B: The correlation model based on expert relationship network;
C: The correlation model based on expert metadata;
D: The model of expert ranking based on listwise with associated features;

The evaluation indexes are MAP, P@5, NDCG@5. In the case of the same data set, experimental results are shown in Fig. 1.

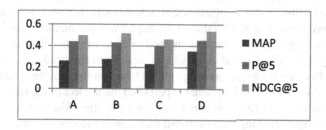

Fig. 1. Experimental results

That can be seen in Fig. 1. The value MAP and NDCG@5 of B is higher than A by nearly 2 percentage points. It illustrates that the correlation model based on expert relationship network has great influence on the ranking of experts. The three evaluation indexes of D were significantly higher than those of A, B, C. It illustrates that three kinds of correlation relationship have contributed to improve the experts ranking, and Expert-ListNet algorithm can be a good fusion of these three kinds of correlation, and effectively improves the ranking of experts.

3.3   The Experimental Comparison of Different Methods of Expert Ranking Methods

In order to test the effect of the model of expert ranking based on list with associated features, it is compared with three kinds of sorting methods which are the ranking method of document model, the ranking method of co-occurrence social network model and the expert ranking method based on listwise.

The evaluation indexes are MRR, MAP, P@5, P@10, NDCG@5 in the experiment. In the same data set, the experimental results are shown in Table 2.

Table 2. Experimental results of different ranking methods

	The document model	The co-occurrence social network model	The expert ranking method based on listwise	The method of expert ranking based on listwise with associated features
MRR	0.6274	0.6554	0.6722	0.7151
MAP	0.2589	0.2617	0.3150	0.3506
P@5	0.4135	0.4200	0.4156	0.4512
P@10	0.3852	0.4013	0.3968	0.4211
NDCG@5	0.4821	0.5145	0.5110	0.5369

As can be seen from the above table, the MAP value of the social network model is improved by nearly 3 percentage points compared with the document model. It illustrates that the introduction of the correlation characteristics of expert can improve expert ranking performance. The MAP value of the expert ranking method based on listwise is significantly higher than the ranking method of document model, which shows that the expert ranking method based on listwise is superior to the ranking method of document model. Because the expert ranking method based on listwise breaks the limitation of the correlation of absolute individual experts and the neglect of the relative relationship of the whole expert list. Compared with the other three kinds of sorting method, the method of expert ranking listwise based on list with associated features is the best, and the evaluation indexes were improved. The validity and superiority of this method are proved.

4   Conclusion

In view of the traditional expert's ranking methods less consider the correlation relationship of experts and the relative relationship among expert list, the method of expert ranking based on listwise with associated features is proposed. Experimental results show that this method improves the accuracy of the expert ranking.

References

1. MacDonald, C., He, B., Plachouras, V., Ounis, I.: University of Glasgow at TREC 2005:experiments in terabyte and enterprise tracks with terrier. In: Proceeding of the 14th Text REtrieval Conference(TREC2005), Gaithersburg, USA, pp. 1–14 (2005)
2. Fu, Y., Yu, W., Li, Y., et al.: THUIR at TREC 2005. enterprise track. In: Proceedings of the 14th Text Etrieval Conference (TREC 2005), Gaithersburg, USA, pp. 1–8 (2005)
3. Balog, K., Azzopardi, L., de Rijke, M.: A language modeling framework for expert finding. J. Inf. Process. Manage. **45**(1), 1–19 (2009)
4. Fang, Y., Si, L.: Discriminative models of integrating document evidence and document-candidate associations for expert search. In: Proceeding of the 33rd International ACM SIGIR, Geneva, Switzerland, pp. 683–690 (2010)
5. Li, N., Ning, K., Zhang, L., Wang, Y.: Expert ranking on the enterprise microblogging based on the PageRank algorithm. In: Asia-Pacific Services Computing Conference, Guilin, China, pp. 345–349 (2012)
6. Chen, J.-M., Liu, J., Huang, Y.-L., Lu, M.: Efficient TOP-K support documents for expert search using relationship. In: A Social Network. Proceedings of the 2011 International Conference on Machine Learning and Cybernetics, Guilin, China, pp.1479–1484 (2011)
7. Liu, Q., Li, S.: Word similarity computing based on how-net. In: Computational Linguistics and Chinese Information Processing, pp. 31(7), Taiwan, China, pp. 59–76 (2002)
8. Chen, C., Peng, B., Yan, H., Wang, J.: A term co-occurrence algorithm and the effect of co-occurrence terms on result ranking for information retrieval. J. Tsinghua Univ(Sci&Tech). **45**(S1), 1857–1860 (2005)
9. Cao, Z., Qin, T., Liu, T.-Y., Tsai, M.-F., Li, H.: Learning to rank: from pairwise approach to listwise approach. In: Twenty-Fourth International Conference on Machine Learning, ICML 2007, pp. 129–136. Oregon State University in Corvallis, Oregon (2007)
10. Fang, Y., Si, L., Mathur, A.P.: Discriminative probabilistic models for expert search in heterogeneous information sources. J. Inf. Retrieval. **14**(2), 158–177 (2011)

Generating Triples Based on Dependency Parsing for Contradiction Detection

Luyang Li, Bing Qin[✉], and Ting Liu

Research Center for Social Computing and Information Retrieval,
Harbin Institute of Technology, Harbin, China
{lyli,qinb,tliu}@ir.hit.edu.cn

Abstract. Contradiction detection is a task to detect the contradictory relation between two texts. In the social media, the phenomenon of contradictory descriptions of the same event is common and harmful. It is urgent to detect contradictory texts. Previous methods on detecting contradiction are mostly deriving features from shallow semantic representations like predicate-argument structures. They meet a problem of the low coverage of contradiction. We propose a joint method to extract more contradiction pairs. We utilize dependency parsing tree to generate tripes (dp-triple) which represent semantic information of the text. The dp-triple extraction method extract more contradiction pairs than present shallow semantic extraction methods like open IE or SRL. Due to the coverage limitation of triples, we also derive features from the context of the matching words between texts as backup. We demonstrate the joint method is effective in detecting contradiction. In predicting stage, we use a unsupervised method to detect contradiction relation and achieve a better performance than the state of the art method.

1 Introduction

Contradiction is also called conflicting data which cannot be true at the same time [6]. Contradiction detection (called CD for short) is a foundational task of text understanding [3]. The technology of detecting contradiction can be used in some other natural language processing tasks like claim-consistency detection which is to find the inconsistent claims or data of the politicians or scientific paper. As the rumors are rampant in the social network, a contradiction detection system is useful to identify the rumors which are contradictory with the fact.

Previous researchers take the task as a classification problem. They derive features from certain semantic representations and mostly use a supervised machine learning method. There is a challenge in the task which is finding a good semantic extraction method to extract semantic information from sentences and help to detect contradiction. SRL or relation extraction methods are some shallow semantic representation methods which extract triples to represent the text. However, these methods miss some important information. For instances, some appositive parses exist "isA" relation, and some complements also can combine with predicates prior to the complement to form a triple, etc.

© Springer Science+Business Media Singapore 2015
X. Zhang et al. (Eds.): SMP 2015, CCIS 568, pp. 200–208, 2015.
DOI: 10.1007/978-981-10-0080-5_19

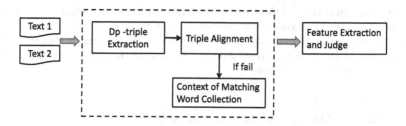

Fig. 1. The architecture of the contradiction detection system.

We present a joint method to solve the problem in a rule-based framework as in Fig. 1. Through the analysis of the corpus, we proposed a dependency parsing based method to extract triples which we called dp-triple extraction. By computing the similarity of the triples between pair of texts, we obtain likely-contradictory pair which maybe contain contradiction. If there is no likely-contradictory pair, we collect the context of matching words by a window. We believe the contradiction maybe exist around the matching words in the pair of texts. Finally, the semantic relation is judged based on the features.

2   Joint Method for Dp-Triple Extraction

2.1   Dp-Triple Extraction Method

Some researches are adopting triples to represent information from sentences, such as Semantic Role Labeling (SRL) and relation extraction. In CD task, some approaches also use these tools or techniques to generate triples. We make a comparison among the SRL method [1], open IE method Reverb[2] and our method on extracted triples in Table 1. SRL runs based on dependency parsing, which aims to label the arguments of each predicate in the sentence. While there are many non-predicate triples in the sentence. In Table 1, the triple *(James Clark, acquaintance, Jones' family)* in the first text can hardly be found because of non-predicate and apposition which is implicit.

Relation extraction is to find various predefined semantic relations between pairs of entities. Most extraction approaches can be roughly classified into two categories: knowledge-based approaches and statistical approaches. Knowledge-based approaches rely on the relation indicators from knowledgebase to extract relation triples, that leads to a low coverage. Statistical approaches such as ReVerb use certain templates to screen out triples. The arguments of the predicate are extracting not by parsing, but by principle of proximity, which will make a number of errors. In the Table 1, ReVerb mistakenly extract the triple *(the mail-box, belonged, James Clark)* from the first text, however, the right

---

[1] We use "Semantic Role Labeler" from UIUC http://cogcomp.cs.illinois.edu/demo/srl/?id=14.

[2] ReVerb is available online on http://reverb.cs.washington.edu/.

Table 1. The comparison of three triple extraction methods. The triples are extracted from the sentence *"The car which crashed against the mail-box belonged to James Clark, 68, an acquaintance of James Jones' family"*. The sentence is contradicting with sentence *"Clark is a relative of Jones's"* because of the different relation indicators "acquaintance" and "relative".

Method	Triples
SRL	(The car, crash against, the mail-box)
	(The car, belonged, James Clark)
Reverb	(The car, crash against, the mail-box)
	(The mail-box, belonged, James Clark)
Dp-triple	(The car, crash against, the mail-box)
	(The car, belonged, James Clark)
	(James Clark, be, 68)
	(James Clark, be, an acquaintance of James Jones' family)
	(James Clark, acquaintance, Jones' family)

triple of predicate "belong" is *(The car, belonged, James Clark)* which argument "the car" is at a long distance from the predicate "belong".

According to the above-mentioned weakness, we proposed a approach to extract the triples based on the dependency parsing. Based on the analysis of the sentence, we categorize triples into two types "Relation of entities or NPs" and "Status of entities or NPs".

Relation of Entities or NPs. Some information content are in the shape of relations of entities or NPs existing in the sentence. The relation indicators with entities or NPs have multiple grammatical relationships in sentences. We list these grammatical relationships and rules of collection to extract the triples.

Predicate-Type: In this type, the relation indicators act as a predicate in the sentence. As we know, in dependency parsing,the dependencies are all binary relations: a grammatical relation holds between a governor (also known as a regent or a head) and a dependent. Around this predicate, there are the followed combinations of grammatical relationships that can be extracted triples: subject and object; subject and preposition; passive subject and preposition.

Possessive-Type: There are two forms in the possessive-type phenomenon, which can simply be described as "s"-form and "of"-form. The "s"-form use "s" to describe the possessive relation of people, while the "of"-form use "of" to describe the possessive relation of thing.

Appositive-Type: In the grammatical relationship appositive, the entities (or NPs) exist "be" (become) relation with each other. The relation is implicit which the relation indicator is absent, however, we can recognize the phenomenon by dependency parsing and completion the relation in the triple.

Status of Entities or NPs. Some information content describe the status of entities or NPs in the sentence. The words represent status can abstractly classify in the followed two types based on the grammatical relationships.

Complement-Type: Some status words act complement as a grammatical item. The combination of grammatical relationships subject and complement which both containing this status word can be extracted triples.

Passive-Auxiliary-Type: Some passive sentences express the status of entities. A passive auxiliary of a clause is a non-main verb of the clause which contains the passive information. The combination of passive subject and passive auxiliary will be extracted triples.

Scope of the Negative Word. Negation is an important problem in the CD task. The texts which express negation meanings, in general, contain negative words (like not) or opposite words (like refuse) to express a opposite meaning of the subsequent predicate or clause, like the following example *"Bill refuse to accept the award"* and *"Bill doesn't drive a car"*

To resolve the negation problem, we will face two subproblems which are negative words (and opposite words) detection and the scope of the negative words recognition. The negative and opposite words detection is a research problem, while we summarize a word list including common words of this kind and have a good coverage on the development set. Recognize the scope is also an important problem which helps to detect the negative triples. The negative word and its following information in the scope can form different syntactic structures. We classify them into two types.

Clause-Type: In this type, the negative words are verbs or preposition with negative or opposite meaning for the following negative scope. The negative scope in this case is the clause followed the negative words. Based on the dependency parsing, we can extract the negative triples from the corresponding syntactic structures. In the first case of clause-type, that negative words are verbs, the negative scope servers as complement in the sentence. In the second case of clause-type, that the preposition "without" guide a clause, the syntactic structure prepositional clause can help to extract negative triples which with the tag "prepc_without" in Stanford Parser.

Predicate-Type: When the negative word is an adverb which acts on the adjacent predicate, we can seize the predicate by dependency parsing. In the Stanford Parser, some negative words with its predicate can be tagged as "neg" structure and others will be detected from the structure "advmod".

The note about the predicate-type is that we do not deal with the last case of predicate-type in the Table 2, which is corresponding to the negative words nouns (nobody, nothing, none, etc.). These negative words are with negative scopes, and act as components in the triples. That will bring more problems in the follow-up steps, triple alignment and contradiction decision. More detailed studies will be done in our next work.

Table 2. Negative words with corresponding scopes.

Type	POS	Words	Syntactic structure of the scope
Clause-type	verb	forget, deny, refuse, reject	ccomp/xcomp
	preposition	without	prepc_without
	adjective	no, unable, incorrect	ccomp/xcomp
Predicate-type	adverb	not, never, hardly	neg/advmod
	noun	nobody, nothing, none	

2.2 Triple Alignment

After dp-triple extraction, multiple triples have been extracted from each text. The alignment of the triples between pair of texts is needed. We treat the alignment of triples as relevance ranking problem. Through analysis of contradiction corpus, we find that the contradictory phenomena exist in pair of triples with similar words. We use t stand for the triple which form is like $< argument_1, relation, argument_2 >$. If $T^T = \{t_1^T, t_2^T, ..., t_i^T, .., t_m^T\}$ and $T^H = \{t_1^H, t_2^H, ..., t_j^H, .., t_n^H\}$ stand for the triple sets of two texts (name as T and H), our aim is to find the most similar triple pair. The objective function is defined as following.

$$F(T^T, T^H) = max_{(t_i^T \in T^T, t_j^H \in T^H)} Sim(t_i^T, t_j^H) \tag{1}$$

The similarity between triples can be computed as Eq. 2 with the condition that any two elements in the total three elements must be more than α separately. Namely, it is the summation of $sim(argument_1^T, argument_1^H)$, $sim(arargumentg_2^T, argument_2^H)$ and $sim(relation^T, relation^H)$. Here, we set an empirical threshold value α as 0.5.

$$Sim(t_i^T, t_j^H) = \sum_{p \in t_i^T, q \in t_j^H} sim(p, q) \tag{2}$$

The similarity between two elements in the pair of triples is computing as Eq. 3.

$$sim(p, q) = \frac{Number\ of\ mapping\ words\ between\ p\ and\ q}{Number\ of\ words\ in\ p} \tag{3}$$

Based on similarity computing, the triple pairs from two text will be ranked. Then, each triple from the first text will have a most relevant triple from the second text and vice versa. Finally, we use Eq. 1 to get the most similar triple pair to derive features in the next stage.

2.3 Context of Matching Word Collection

Some pairs of texts share different syntactic structures, so that they do not share a similar triple. We collect the context of the matching words as a backup to derive features. Actually, we think the mismatches contain contradictory phenomena most likely around the matching words in a text. We use a sliding window to obtain the mismatch words around matching words. The window size is 2 in the experiment.

3   Features for Contradiction Detection

In this section, we define seven features to capture contradictory phenomena. They can be concluded to three types.

Inconsistent. The inconsistent of entities in the pair of triples can lead to contradiction. We use Standford CoreNLP to recognize entities from text. Three types of entities are important in the task which are number, person and location. These three kind of entity mismatches can indicate contradiction.

Antonymy. Antonyms are a good cue for contradiction. We find antonyms by using WordNet.

Negation. If one text in the pair contain negation words before aligned content, it almost surely be a contradiction. Negation words are also good indicator of contradiction. In our method, the condition that the feature is set to 1 is negation word occurs before aligned triples or in the context of matching words.

In the classifying stage, we classify the relation of the pair of texts by the rule. If any feature in the feature set is 1, the label of the sample is set to contradiction.

4   Experiment and Analysis

4.1   Data Sets

On account of the lack of benchmark data sets in social media, we use three data sets which are public benchmark data sets from Recognition Text Entailment (RTE) task [1,7,8]. In the original data sets in RTE task, there are three classes which are three semantic relations, contradiction, entailment and unknown (viewed as uncorrelated). We merge the last two classes of data into one, then we get the data sets as showed in Table 3. The distribution of data in two classes is unbalanced. The proportion of contradiction data in the whole data is from 10 % to 15 % (Table 4).

Table 3. Statistics of the three RTE test data sets.

Data set	Contradiction	Non-contradiction	Total
RTE-3 Test	72	728	800
RTE-4 Test	150	850	1,000
RTE-5 Test	90	510	600

Table 4. Experiment results on RTE-3 pilot, RTE-4 test and RTE-5 test data sets of contradiction detection.

Model	RTE-3 Pilot			RTE-4 Test			RTE-5 Test		
	P	R	F1	P	R	F1	P	R	F1
De Marneffes method	22.95	19.44	21.04	–	–	–	–	–	–
BLUE system	–	–	–	41.67	10.00	16.13	42.86	6.67	11.54
Average result	10.72	11.69	11.18	25.26	13.47	13.63	26.40	13.70	14.79
Two-stage method	14.00	19.44	16.27	23.00	22.67	22.82	21.14	28.89	24.40
Joint method (our system)	18.55	31.94	**23.47**	24.30	23.33	**23.81**	27.13	38.89	**31.96**

4.2   Results and Analysis

Baseline methods:

(1) *DeMarneffe's method*: De Marneffe et al. [6] first merge RTE three-way data sets to do the contradiction detection researching. They adopt dependency graphs produced by the Stanford parser as a semantic representation. The system do an alignment between graphs and extract features based on mismatches between texts. They apply logistic regression to do the two-way classification.

(2) *BLUE system*: The BLUE system results are from Boeing's team [2] which run the system on RTE-4 and RTE-5. They use the logical inference approach which adopts logic-based representations of the texts.

(3) *Average result*: The average result is from the submitted systems for three-way task at three data sets.

(4) *Two − stage method*: The two-stage method is the work of Pham et al. [12]. They use SRL tool and ReVerb to extract predicate-argument triples as shallow semantic representation to represent the text. The system also is a rule-based system.

Figure shows the results of the contradiction detection experiment. Our method outperform the baseline methods at all three data sets. The De Marneffes method is a supervised method and is strong at the RTE-3 test set. As an unsupervised method, the two-stage method do not beat the former, but we outperform the supervised method at the F1 value. We also have good performance at the RTE-4 and RTE-5 test sets. Comparing with the De Marneffes method, we obtain a lower precision and a higher recall value. It shows our method can find more contradiction phenomena.

5   Related Work

Contradiction is a kind of semantic relation to pairs of texts. Condoravdi et al. first argue the importance of handling contradiction in text understanding, however, they use a strict logical definition of contradiction phenomena and have no empirical results. Harabagiu et al. focus on the contradiction only caused by negation or formed by paraphrase [9]. Ritter et al. reflect the prevalence of seeming contradictions and the paucity of genuine contradictions in their corpus. For example, "Mozart was born in Salzburg" does not contradict "Mozart was born in Austria" due to meronyms [13]. Hashimoto et al. present a kind of sematic orientation, namely excitatory or inhibitory, like the words "cause" and "ruin". They argue excitation is useful for extracting antonyms [10]. As a kind of relation in entailment recognition problem, contradiction has been studied continually in the Recognizing Textual Entailment Challenges (RTE) [1,4,5,8,11]. The approaches participating in the evaluation mostly concentrate on feature engineering.

6   Conclusion

In this paper, we present a joint method to extract contradictory triples for detect contradiction detection. Current approaches almost are based on the shallow semantic representation methods which meet the low coverage limitation. We address the issue by exploiting a system to extract contradiction triples based on the dependency parsing tree. More contradictory phenomena can be found by our system than shallow semantic extraction methods like ReVerb or SRL tools. On account of the existence of non-alignment of triples, we also adopt the context of the matching words as backup to derive contradiction features. We apply an unsupervised method to predict the label of data. Experimental results show that our system outperforms the state of the art methods.

References

1. Bentivogli, L., Dagan, I., Dang, H.T., Giampiccolo, D., Magnini, B.: The fifth pascal recognizing textual entailment challenge. Proc. TAC **9**, 14–24 (2009)
2. Clark, P., Harrison, P.: Recognizing textual entailment with logical inference. In: Text Analysis Conference (TAC 2008) Workshop-RTE-4 Track, National Institute of Standards and Technology (NIST). Citeseer (2008)
3. Condoravdi, C., Crouch, D., De Paiva, V., Stolle, R., Bobrow, D.G.: Entailment, intensionality and text understanding. In: Proceedings of the HLT-NAACL 2003 Workshop on Text Meaning, vol. 9, pp. 38–45. Association for Computational Linguistics (2003)
4. Dagan, I., Dolan, B., Magnini, B., Roth, D.: Recognizing textual entailment: Rational, evaluation and approaches-erratum. Nat. Lang. Eng. **16**(01), 105–105 (2010)
5. Dagan, I., Glickman, O., Magnini, B.: The PASCAL recognising textual entailment challenge. In: Quiñonero-Candela, J., Dagan, I., Magnini, B., d'Alché-Buc, F. (eds.) MLCW 2005. LNCS (LNAI), vol. 3944, pp. 177–190. Springer, Heidelberg (2006)
6. De Marneffe, M.C., Rafferty, A.N., Manning, C.D.: Finding contradictions in text. ACL **8**, 1039–1047 (2008)
7. Giampiccolo, D., Dang, H.T., Magnini, B., Dagan, I., Cabrio, E., Dolan, B.: The fourth pascal recognizing textual entailment challenge. In: Proceedings of the First Text Analysis Conference (TAC 2008). Citeseer (2009)

8. Giampiccolo, D., Magnini, B., Dagan, I., Dolan, B.: The third pascal recognizing textual entailment challenge. In: Proceedings of the ACL-PASCAL Workshop on Textual Entailment and Paraphrasing, pp. 1–9. Association for Computational Linguistics (2007)

9. Harabagiu, S., Hickl, A., Lacatusu, F.: Negation, contrast and contradiction in text processing. AAAI **6**, 755–762 (2006)

10. Hashimoto, C., Torisawa, K., De Saeger, S., Oh, J.H., Kazama, J.: Excitatory or inhibitory: a new semantic orientation extracts contradiction and causality from the web. In: Proceedings of the 2012 Joint Conference on Empirical Methods in Natural Language Processing and Computational Natural Language Learning, pp. 619–630. Association for Computational Linguistics (2012)

11. Marelli, M., Bentivogli, L., Baroni, M., Bernardi, R., Menini, S., Zamparelli, R.: Semeval-2014 task 1: evaluation of compositional distributional semantic models on full sentences through semantic relatedness and textual entailment. SemEval-2014 (2014)

12. Pham, M.Q.N., Nguyen, M.L., Shimazu, A.: Using shallow semantic parsing and relation extraction for finding contradiction in text (2013)

13. Ritter, A., Downey, D., Soderland, S., Etzioni, O.: It's a contradiction–no, it's not: a case study using functional relations. In: Proceedings of the Conference on Empirical Methods in Natural Language Processing, pp. 11–20. Association for Computational Linguistics (2008)

Improving Conversational Spoken Language Machine Translation via Pronoun Recovery

Yanlin Hu[✉], Heyan Huang, Ping Jian, and Yuhang Guo

Beijing Engineering Research Center of High Volume Language Information
Processing and Cloud Computing Applications,
Department of Computer Science and Technology,
Beijing Institute of Technology, Beijing 100081, China
{yanlinhu,hhy63,pjian,guoyuhang}@bit.edu.cn

Abstract. Machine translation for social communication is necessary in daily life. However, spoken language translation faces many challenges especially in the translation of zero pronouns which is absent in the source language but appear in the target language. Dropping of pronouns severely affects the machine translation from pronoun dropped language such as Chinese to other languages. This phenomenon occurs more frequently in the conversational spoken language. In order to solve this problem, we insert the position of missing pronouns into the source side, then we use the word alignment method to filter the pronouns in order to pick up the pronouns which are really helpful for the machine translation. We achieve improvement on the translation of chat, message and telephone conversational corpus.

Keywords: Pronouns recovery · Word alignment · Conversational spoken language · Machine translation

1 Introduction

Conversational spoken language machine translation is very important in social media area. However, people often intend to omit a lot of compositions which can be inferred from the conversation but hard to be understand by the machine. This phenomenon brings more challenges to the translation. In this paper, we focus on the pronoun drops phenomenon. Pronoun drops happens frequently in conversational languages especially in Chinese. Because Chinese is a kind of pro-drop language whose pronouns are frequently or regularly dropped when they are pragmatically inferable [1]. We found that pro-drop phenomenon occurs once in every two sentences on average of Chinese SMS/Chat corpus. Furthermore, pronouns are difficult to be recovered on the target side when translating pronoun drop language into languages such as English where pronouns are regularly retained. Since it's hard to generate pronouns from nothing.

In addition, pronoun recovery in conversational spoken language machine translation faces more challenges than written languages. As it gives less information than written corpus such as news. We do the statistical analysis on the

© Springer Science+Business Media Singapore 2015
X. Zhang et al. (Eds.): SMP 2015, CCIS 568, pp. 209–216, 2015.
DOI: 10.1007/978-981-10-0080-5_20

XinHua corpus and the People daily news in the year 2000, the averages of words per sentence are 55.9 and 51.5 respectively. However, the Broad Operational Language Translation (BOLT)[1] corpus including chat and telephone message conversations has about 8 words per sentence which often need more information to be added for the translation. Moreover, the syntactic structures in conversational corpus are weaker than those in written corpus, which makes it more difficult to make use of the syntactic information to recover the missing pronouns [2]. In a word, pronoun recovery for spoken language machine translation faces a big challenge.

As we all know, in phrase-based translation, the phrase-based system has the ability to learn pronouns as a part of larger phrases. If the learned phrases include pronouns on the target side that are dropped from source side, the system could insert pronouns even when they are missing from the source side [1]. However, it inserts pronouns probabilisticly. There are still a lot of translation cases that lost their pronouns, such as the second and third sentences in Table 1.

Table 1. An example of conversation translation. The first column shows the source side of the conversation with many dropped pronouns. Square tags in the source side shows the missing pronouns. The second column shows the reference side. And the baseline translation is the statistical machine translation system of Moses [3]

Source	Reference	Baseline Translation
A:你 又 在 出差 ？	A:Are you going on a business trip again?	A:you are you on a trip again? business trip ?
B:没有 □ 在 深圳 呢	B:No, **I** am in Shenzhen.	B:there is no in Shenzhen?
B:□ 后天 出差	B:**I** am going on a business trip. the day after tomorrow.	B:the day after tomorrow on a business trip
A:呵呵 ， □ 真 辛苦	A:Hoho, it's really tough for **you**.	A:hehe , that 's really hard
A:□ 到处 跑	A:Running everywhere.	A:run everywhere

There are several different strategies focused on the pronoun recovery. Pronoun resolution is one way to get the exactly right pronouns in the language using context information [4,5]. However, in Cuillou's work [5] they didn't study the pronoun drop phenomenon and both of the work didn't concern the spoken machine translation. Recovering the zero pronouns Directly in the machine translation brings more noises.

Another helpful approach is explicitly inserting missing words. This helps to make up the gap of structure between the different languages. [1,6,7] employed preprocessing techniques to add some empty categories which contained pronouns position tags into the source text, and achieved improvement on written language machine translation.

---

[1] This corpus comes from the DARPA Broad Operational Language Translation (BOLT) Program which includes message, chat,and telephone conversation parallel data sets The website is https://www.ldc.upenn.edu/sites/www.ldc.upenn.edu/files/bolt_1.pdf.

However, we find that inserting empty categories into source text directly doesn't work well for the conversational language translation between the non-pronoun-drop language and the pronoun-drop languages. This is because that empty categories are only involved in the source side to label the missing part such as pronouns while the target side may not need any pronouns to be involved. This phenomenon is quite common in imperative sentences and exclamatory sentences which doesn't need pronoun to be added on the target side.

In this paper, we propose a new way to improve conversational language machine translation via pronoun recovery. We take the pronouns recovery problem as an sequence tagging problem. Words are tagged to indicate whether there needs an pronoun. Moreover, in order to get better pronoun recovery which is more suitable for the conversation machine translation, we utilize the word alignment method to filter the pronoun tags and choose the useful ones to help improve the performance of the translation.

The rest of this paper is organized as follows. Firstly, we describe a kind of traditional method to get the pronoun position. Secondly, we introduce a new method to get more helpful pronoun tags based on parallel corpus. Finally, we involve the new pronoun tags into the hierarchy based translation model, which shows improvement on the conversational spoken language corpus.

2   Pronouns Recovery

Regarding pronouns recovery as sequence labeling problem is a good way to tag the zero pronoun position. However, this method has shortcomings as well. For example, it sometimes brings unwanted translation. In this section, we firstly tag pronouns based on Conditional random fields method. And then we introduce a new way to solve the shortcomings.

2.1   Pronoun Tagging Based on Conditional Random Fields

The missing pronouns belong to the empty categories which are kinds of element in a parse tree that does not have a corresponding surface word. Chinese Tree Bank (CTB) [8] has different kinds of empty categories, but we only focus on the "*pro*" tagged sentences which show the position of missing pronouns. Moreover we treat all the pronouns as one category, in this way we avoid the data sparse problem. We extract the appropriate sentences from CTB. Then we build a simple conditional random fields [9] model on CTB corpus to recover pronouns. The part-of-speech and lexical features are employed to train the pronouns tagging model. Then we use the trained model on the BOLT corpus to tag the missing pronouns. Finally, the tagged corpus are utilized to train translation models. We finally recover serval pronouns in the target side as the bold shows in the third column of the Table 2.

However, the improvement of translation doesn't seem significant. Sometimes even got some decline than the non pronoun tagged translation. We analyze the translation result, and found that some "*pro*" could be translated into

Table 2. An example of pronoun tagged conversation translation. The first column shows the source side of conversation with the tagged pronouns. *pro* in the source side shows the position of the missing pronouns. The second column shows the reference side. And the pronoun tagged translation is the general translation of pronoun tagging sentences

Source	Reference	Pronoun tagged Translation
A:你 又 在 出差 ？	A:Are you going on a business trip again?	A:you are you on a business trip?
B:没有 ***pro*** 在 深圳 呢	B:No, I am in Shenzhen.	B:**I** don't have it in Shenzhen ?
B:***pro*** 后天 出差	B:I am going on a business trip the day after tomorrow.	B: **I** will be on a business trip the day after tomorrow
A:呵呵 ， ***pro*** 真 辛苦	A:Hoho, it's really tough for **I**.	A:hehe , **that** 's really hard
A:***pro*** 到处 跑	A:Running everywhere.	A:**I** run everywhere

the pronouns, but a majority of "*pro*" could not. Therefore, there are many pronoun tagged words that bring some unnecessary noise translations, and this may lead the decline.

2.2 Pronouns Recovery with Filtering

In order to recover pronouns in the target side without reducing the performance of the machine translation, we introduce a filter to pick up the position of pronouns which can truly help the translation.

Firstly, we use the pronoun tagged conversational data to make the word alignment [10]. We assume that the tags which are aligned to the pronouns of target side such as "you", "I", "he", etc., are intended to be the true pronoun tags that will be useful for translation. Under this assumption, we retain the pronoun tags which are aligned to pronouns and filter out the others. Then we use these filtered data as our new training set for pronouns recovery which helps us find more appropriate tags for the translation. And we get improvements in translation.

In order to verify the assumption, we compute the lexical translation table on the internet chat (CHT) corpus from BOLT. Table 3 shows the results of lexical translation table on the CHT corpus with pronoun tagged and CHT corpus with pronoun filtered respectively. The probabilities of pronouns increase. On the contrary, the other words' probabilities decrease.

Table 3. Part of lexical translation table

Word	$P(word| *pro*)$ CHT	$P(word| *pro*)$ filtered CHT
right	0.154	**0.004** ↓
you	0.114	**0.191** ↑
it	0.162	**0.287** ↑

3 Experiments

3.1 Experimental Setup

We used three genres of conversational spoken language corpus on BOLT. They are conversational telephone speech (CTS), message conversation (SMS) and internet chat (CHT) respectively. The details of these corpus are listed in Table 4. For each genre of corpus, the test set and the development set consist of about 5000 sentences, respectively.

All the Chinese data was segmented by the Language Technology Platform (LTP) tools [11]. The baseline system was trained using the BOLT corpus without pronoun tagged. Our experiment used the Hiero [12] system embedded in the Moses [3]. The training data of the language model is the English side of parallel training corpus. And we trained a 7-gram language model with the modified Kneser-Ney smoothing [13]. The same number of GIZA++ [14] iterations was used for all the experiments. Minimum error training [15] was conducted on every system. For each system, the BLEU score [16] was calculated on the test set.

3.2 Result and Discussion

The BLEU scores of different tagging methods for source side data and different corpus are shown in Table 5. All the results in Table 5 marked with "*" are statistically significant with p<0.05 based on 1000 iterations of paired bootstrap resampling [17]. As we can see that the traditional method involving pronouns directly in the statistic machine translation system receives a slight decline. However, when using the pronouns filtering method, significant improvements achieve on either of chat, message, telephone conversation corpus.

Table 4. Details of the training corpus

Train corpus	Words	Sentences
CTS	989,512	94,685
SMS	29,011	3,648
CHT	833,230	120,920

Table 5. The results of different systems on different corpus. Baseline is the general hierarchy based statistical machine translation system. Baseline+tag is short for the pronoun tagging system and Baseline+fTag is short for the system with filtered tagging pronouns

System	CHT	SMS	CTS
Baseline	14.72	9.24	21.17
Baseline+tag	14.70	8.56	20.09
Baseline+fTag	14.93*	9.39*	21.22*

Table 6 shows four examples of translations on the test sets. The table includes the translation results from the system trained without pronoun filtered, and the system trained with pronoun filtered. It shows that some pronoun tags hurt the translations with involving addition translated words or pronouns that are not necessary in the target side.

Table 6. Three examples of translations on the test sets

Source w/all nulls	*pro* 对吧?
Source w/ filtered nulls	对吧?
Reference	Right?
System trained w/all nulls	Right, Right?
System trained w/ filtered nulls	Right?
Source w/all nulls	*pro* 没关系，我发你东西了，*pro*快接收
Source wo/nulls	*pro* 没关系，我发你东西了，快接收
Reference	It doesn't matter, I have sent you something, Quick, accept it.
System trained w/all nulls	It doesn't matter, I sent it to your things, almost receive.
System trained w/ filtered nulls	It doesn't matter, I sent it to your things, it's almost accept.
Source w/all nulls	*pro* 不够
Source wo/nulls	不够
Reference	Not Enough
System trained w/all nulls	I don't have enough.
System trained w/ filtered nulls	Not enough.

4 Conclusions and Future Work

Dropping of pronouns brings challenge to the machine translation. Our experiment shows that the traditional method which inserts empty categories directly doesn't work well in this area. In this paper, we investigate a new way to pick up more appropriate pronouns to ensure reasonable pronoun recovery in the target side.

Using the new way to get the positions has several advantages. On the one hand, we put our focus on the true pronoun dropped positions which actually helps to go across the structure gaps so that the tagged positions can be aligned to pronouns on larger probability. On the other hand, this method gets rid of some redundant words which may lead to more unrelated words involved in the translation results. In this paper, we improved the conversational spoken language translation via pronouns tagging.

In the future work, we will consider integrating the pronoun tags into the decode process, so that it can benefits from interacting with other compositions. We will also consider involving pronouns inference process into the machine translation system. No matter how hard it is, it's worth getting the correct missing pronouns which contribute greatly to the conversational spoken language machine translation.

Acknowledgments. This work is supported by the National Basic Research Program of China (973 Program, Grant No. 2013CB329303), the National Natural Science Foundation of China (Grant No. 61132009, 61202244) and Beijing Institute of Technology Research Fund Program for Young Scholars.

References

1. Chung, T., Gildea, D.: Effects of empty categories on machine translation. In: Proceedings of the 2010 Conference on Empirical Methods in Natural Language Processing (2010)
2. Wang, H., Gao, W., Li, S.: Speech machine translation research review. Comput. Sci. **5**, 47–50 (1998)
3. Hoang, H., Birch, A., Callison-Burch, C., Zens, R., Federico, M., Bertoldi, N., Dyer, C., Cowan, B., Shen, W., Moran, C.: Moses: open source toolkit for statistical machine translation. Proc. Assoc. Comput. Linguist. **9**(1), 177–180 (2007)
4. Chen, C., Ng, V.: Chinese zero pronoun resolution: some recent advances. In: Proceedings of the 2013 Conference on Empirical Methods in Natural Language Processing, pp. 1360–1365 (2013)
5. Guillou, L.: Improving pronoun translation for statistical machine translation. In: Proceedings of the Student Research Workshop at the 13th Conference of the European Chapter of the Association for Computational Linguistics, pp. 1–10 (2012)
6. Yamada, K., Knight, K.: A syntax-based statistical translation model. In: Proceedings of the 39th Annual Meeting on Association for Computational Linguistics, pp. 523–530 (2001)
7. Xiang, B., Luo, X., Zhou, B.: Enlisting the ghost: modeling empty categories for machine translation. In: ACL, pp. 822–831 (2013)
8. Xue, N., et al.: Chinese Treebank 8.0 LDC2013T21. Linguistic Data Consortium, Philadelphia (2013)
9. Lafferty, J., Mccallum, A., Pereira, F.: Conditional random fields: probabilistic models for segmenting and labeling sequence data. In: Proceedings of the 18th International Conference on Machine Learning 2001 (ICML 2001), pp. 282–289 (2001)
10. Vogel, S., Ney, H., Tillmann, C.: HMM-based word alignment in statistical translation. In: Proceedings of the 16th Conference on Computational Linguistics, vol. 2, pp. 836–841 (1996)
11. Che, W., Li, Z., Liu, T.: LTP: a chinese language technology platform. In: Proceedings of the 23rd International Conference on Computational Linguistics, pp. 13–16 (2010)
12. Chiang, D.: A hierarchical phrase-based model for statistical machine translation. In: Proceedings of the 43rd Annual Meeting of the Association for Computational Linguistics, pp. 263–270 (2005)
13. Chen, S.F., Goodman, J.: An empirical study of smoothing techniques for language modeling. Comput. Speech Lang. **13**(4), 359–393 (1999)
14. Och, F.J., Ney, H.: A systematic comparison of various statistical alignment models. Comput. Linguist. **29**(1), 19–51 (2003)

15. Och, F.J.: Minimum error rate training in statistical machine translation. In: Proceedings of the 41th Annual Meeting on Association for Computational Linguistics, vol. 32, no. 17, pp. 701–711 (2003)
16. Papineni, K., Roukos, S., Ward, T., Zhu, W.J.: BLEU: a method for automatic evaluation of machine translation. In: Proceedings of the 40th Annual Meeting on Association for Computational Linguistics, pp. 311–318 (2002)
17. Koehn, P.: Statistical significance tests for machine translation evaluation. In: EMNLP, pp. 388–395. Citeseer (2004)

Analyzing the Segmentation Granularity of RTB Advertising Markets: A Computational Experiment Approach

Rui Qin[1,2]($\boxtimes$), Yong Yuan[1,2], Feiyue Wang[1,2], and Juanjuan Li[1,2]

[1] The State Key Laboratory of Management and Control for Complex Systems, Institute of Automation, Chinese Academy of Sciences, Beijing, China
{rui.qin,yong.yuan,feiyue.wang,juanjuan.li}@ia.ac.cn
[2] Qingdao Academy of Intelligent Industries, Qingdao, China

Abstract. Real Time Bidding (RTB) is an emerging business model of online computational advertising with the rise of Internet and big data. It can help advertisers achieve the precision marketing through evolving the traditional business logic from buying ad-impressions to directly buying the matched target audiences. As an important part of RTB markets, Demand Side Platforms (DSPs) play a critical role in matching advertisers with their target audiences via market segmentation, and their segmentation strategies (especially the choice of granularity) have key influences in improving the efficiency of RTB markets. This paper studied DSPs' strategies for market segmentation, and established a selection model of the granularity for segmenting RTB markets. We proposed to validate our model using a computational experiment approach, and the experimental results show that the market segmentation granularity has the potential of improving both the total revenue of all the advertisers and the expected revenue for each advertiser.

Keywords: Real time bidding · Demand side platforms · Market segmentation granularity · Computational experiments · Precision marketing

1 Introduction

With the rapid development of Internet and big data, Real Time Bidding (RTB) has been widely-recognized as the most popular business model of online advertising markets, and an inevitable trend of the sale model for online digital media. RTB can help advertisers to reach their target audiences via real-time, auction-based matching and pricing, and thus display their advertisements to the right person at the right time with lower costs. As such, RTB has the potential of improving the market efficiency as well as advertisers' revenue.

As a central part in the RTB ecosystems, Demand Side Platforms (DSPs) stand on the demand side of this market, serving as an agency making decisions on behalf of their advertisers. Typically, designing efficient bidding algorithms

© Springer Science+Business Media Singapore 2015
X. Zhang et al. (Eds.): SMP 2015, CCIS 568, pp. 217–224, 2015.
DOI: 10.1007/978-981-10-0080-5_21

and effective market segmentation strategies are key decisions for DSPs. In literature, designing bidding algorithms attracted much more interests of researchers. For instance, Ghosh et al. [2] proposed an offline bidding algorithm for DSPs that is suitable for both the full information and the partially observable information settings. Other offline algorithms have been proved to be able to optimize advertisers' bids based on the historical average winning bid prices [6], the predicted winning rates and prices [3], as well as the click-through-rates [5]. Chen et al. [1] designed an online bidding algorithm that can support fine-grained impression valuation and adjust value-based bids dynamically.

As for DSPs' market segmentation strategy, related research is still nonexistent. Via analyzing cookie-based online big data, DSPs can in theory infinitely precisely segment advertisers' target audiences into large numbers of niche markets. In RTB practice, the granularity used by DSPs is a key parameter that can partly determine both the competition degree among advertisers and their valuations of ad impressions. On one hand, fine-grained market segmentation can improve the precision of advertiser-audience matching and also advertisers' values per ad impression. This helps increase the average price of RTB ads. On the other hand, however, with the increasing segmentation granularity, the number of advertisers in each niche market, and also the competition among them, will be reduced. This will decrease the RTB ad price. Obviously, there exists a dilemma for DSPs in choosing the granularity in pursuit of better ad prices and revenue. Therefore, it is extremely urgent to study the market segmentation problems, and provide a feasible market segmentation strategy for DSPs, so as to maximize the marketing effect of advertisers in RTB markets.

In this paper, we aim to preliminarily study the market segmentation problem for DSPs, and define it as an issue of seeking for the best market segmentation granularity within the given alternative granularities. Considering the market complexity, we utilize a computational experiment approach to validate the effectiveness of our proposed model. Experimental results show that the increasing of the market segmentation granularity can greatly improve the total revenue of all the advertisers on the DSP, and thus the marketing effect of RTB advertising.

The rest of the paper is arranged as follows: In Sect. 2, we first introduce our problem briefly, and then establish a model for the choice of market segmentation granularity. In Sect. 3, we propose to use a computational experiment approach to solve our proposed model, and design numerical experiments to validate our model. Section 4 discusses the managerial insights of our findings for DSPs. Section 5 concludes our efforts.

2   The Model

2.1   Problem Statement

In RTB markets, DSPs typically label the target audiences (or users) with various kinds of tags, resulting in a hierarchical structure shown in Fig. 1. For example, user 1 may represents the people who love sports, then user 2(1) and user 2(2) can represent the men who love sports and the women who love sports, respectively.

Obviously, the granularity of market segmentation increases from the top level to the bottom level, and correspondingly the number of users in each niche market decreases, resulting in better matching and targeting for advertisers.

For DSPs, each level corresponds to a market segmentation strategy, and these strategies may lead to different marketing effect for advertisers. Thus, a DSP has to choose the best market segmentation strategy, so as to maximize the marketing effect for all advertisers.

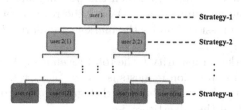

Fig. 1. The user structure in the DSP

2.2   The Model

We consider the case when there is only one DSP in the market, i.e., the winning advertiser will obtain the ad impression. Suppose there are M market segmentation granularities, denoted as $L = \{1, 2, \cdots, M\}$, and correspondingly, the DSP has M market segmentation strategies. Denote the advertisers on the DSP as $U = \{1, 2, \cdots, N\}$, and their budgets as $B = \{B_1, B_2, \cdots, B_N\}$. For a certain time period, there are S ad impression requests, denoted as $Q = \{1, 2, \cdots, S\}$, and for each request $j \in Q$, the reserve price is ρ_j.

For each market granularity $l \in L$, the matching probability between the advertiser $i \in U$ and the ad impression request $j \in Q$ is $\sigma_l(i, j) \in [0, 1]$. The matching probability is a measurement for the matching degree of the ad impression (and also the audience) with the advertiser. $\sigma_l(i, j) = 0$ and $\sigma_l(i, j) = 1$ represent a complete mismatch and a perfect match, respectively. For any given $\alpha \in [0, 1]$, the advertiser bids for the ad impression only if $\sigma_l(i, j) \geq \alpha$ and the remaining budget is sufficient.

Denote the value function of the advertiser $i \in U$ for ad impression j under a market granularity l as $v_i(l, j)$. According to the equilibrium properties of Vickrey auction mechanism, we can assume that the bid for advertiser i for ad impression j is also $v_i(l, j)$. The remaining budget for advertiser i after buying ad impression j is $b_{l,i}(j) = b_{l,i}(j - 1) - c_{l,i}(j - 1)$ for $j = 1, 2, \cdots, S$, where $b_{l,i}(0) = B_i$, and $c_{l,i}(j - 1)$ represents the cost of advertiser i for ad impression $j - 1$ under a market granularity l. We set $c_{l,i}(j - 1) = 0$ if advertiser i does not win in the auction.

Under each market granularity l, the set of advertisers bidding for ad impression request j can be computed as follows

$$U_\alpha(l, j) = \{i | i \in U, \sigma_l(i, j) \geq \alpha, b_{l,i}(j) \geq v_i(l, j)\}, \tag{1}$$

and the advertisers with the highest bid and the second highest bids can be computed by

$$i_0(l, j) = \underset{i \in U_\alpha(l,j)}{\mathrm{argmax}} \, v_i(l, j), i'(l, j) = \underset{i \in U_\alpha(l,j)/i_0(l,j)}{\mathrm{argmax}} \, v_i(l, j). \tag{2}$$

According to the RTB auction mechanism, the advertiser with the highest bid wins the auction, and he or she needs to pay only the second highest bid. Thus, the advertiser $i_0(l, j)$ wins the auction, and the cost is $c_{l,i_0}(j) = \max\{v_{i'(l,j)}, \rho_j\}$.

Assume the revenue of the advertiser from an ad impression is equal to the advertiser's value for the impression, then the winning advertiser i_0 can obtain $v_{i_0}(l, j)$ revenue from ad impression j. Denote the revenue of all advertisers on the DSP from ad impression j under the market granularity l is $r(l, j)$, then we have $r(l, j) = v_{i_0}(l, j)$.

Thus, under market granularity l, the total revenue of the advertisers on the DSP from all the ad impression requests is $g(l) = \sum_{j \in Q} r(l, j)$.

The DSP aims to choose the best market granularity from L to maximize the total revenue of all the advertisers, i.e.,

$$\max_{l \in L} g(l). \tag{3}$$

Solving the above model, we can obtain the optimal market granularity l^*, and the corresponding optimal revenue is $g(l^*)$.

3   The Experiments

Due to the essential complexity of online RTB markets, it is difficult or even impossible to validate our proposed model and strategies with online field experiments. Fortunately, computational experiments [4] can serve as an alternative way. In this section, we will design the experimental environment with the computational experiment approach, to validate our model.

3.1   The Computational Experiments Scenario

Suppose there is only one DSP, and the number of ad impression requests during a certain time period on the DSP is $S = 1000000$. The hierarchical structure of the users corresponding to these impressions is shown in Fig. 2, with three layers. The proportion of these users are $a1 : a2 : b1 : b2 = 1 : 1 : 1 : 1$, and these users are uniformly distributed in the time period. The reserve price of these ad impression requests is $\rho = 2.00$.

According to the three-layer-structure of the users in Fig. 2, the DSP has three feasible market segment strategies, each corresponding to a layer. The three strategies are defined as follows:

(1) Strategy-I: no market segmentation, corresponding to the top layer;
(2) Strategy-II: segmenting the market into two niche markets, corresponding to the middle layer;
(3) Strategy-III: segmenting the market into four niche markets, corresponding to the bottom layer;

Table 1. The targeting audience for the 8 advertisers under different segment strategies

Advertiser	Targeting population			Total budget
	Strategy-I	Strategy-II	Strategy-III	
A	All	a	a1	1500
B	All	b	b1	1300
C	All	a	a2	1000
D	All	b	b2	800
E	All	a	a1	700
F	All	b	b1	600
G	All	a	a2	500
H	All	b	b2	200

To evaluate the three strategies, we construct an experiment with 1 DSP and 8 advertisers. Their targeting audiences under each strategy are given in Table 1. For each strategy, their CPMs for the impressions are fixed.

In RTB practice, advertisers' values of ad impressions typically increase with the accuracy of matching, and so we assume that the CPMs of the advertisers under the three strategies are uniformly distributed in $[2.20, 8.00]$, $[4.00, 15.00]$ and $[7.00, 20.00]$, respectively. Moreover, for each strategy, we assume $CPM(A) > CPM(B) > \cdots > CPM(H)$ for these advertisers' CPMs. This assumption is reasonable since generally the advertiser with a higher CPM under one segmentation strategy also has a higher CPM under others.

3.2 The Experimental Results

In order to obtain general conclusions as for the influence of the segmentation strategy on advertisers' revenues, we conduct 2000 independent experiments, and in each experiment, the process for finding the optimal market segmentation granularity is given in Fig. 3.

With the above solving process, the total revenue and the average total revenue for the advertisers on the DSP, and the average revenue for each advertiser in these 2000 experiments are shown in Fig. 4.

From Fig. 4, we can obtain the following conclusions:

(1) For all the 2000 experiments, the total revenue increases with the market segmentation granularity.
(2) The average total revenue for the three strategies are 6098.659, 9150.510 and 12777.387, respectively. It is obvious that Strategy-III outperforms Strategy-I and Strategy-II (about 101.51 % and 39.64 %), and Strategy-II outperforms Strategy-I about 50.04 %, in terms of the average total revenue.
(3) For each advertisers, the average revenue increases with the market segmentation granularity.

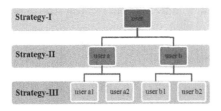

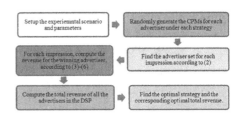

Fig. 2. The three-layer-structure of the users in the DSP

Fig. 3. The process for finding the optimal strategy

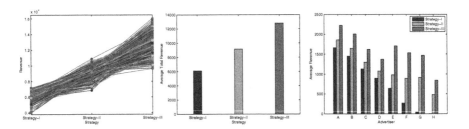

Fig. 4. The total revenue on the DSP, the average total revenue on the DSP, and the average revenue for each advertiser in 2000 experiments

Furthermore, Fig. 5 gives the revenues for the 8 advertisers in the 2000 experiments, and it is obvious that there are 6 cases for the relations of the revenues among the three strategies, as shown in Fig. 6.

In order to make a better comparison of the revenues among the three strategies, we give the percentage of each case for every advertiser in the 2000 experiments, as shown in Fig. 7. From the results, we can see that case-1 occurs with proportions of 58.15 %, 62.55 %, 61.05 %, 64.55 %, 77.45 %, 84.6 %, 73.4 % and 94.8 % for advertiser A–H in all experiments, respectively, which illustrates that Strategy-III outperforms the other two strategies in most situations.

3.3   Analysis of the Experimental Results

From the above experimental results, we can obtain the following findings:

(1) With the increasing of market segmentation granularity, both the total revenue and the average total revenue of all the advertisers on the DSP, as well as the average revenues of each advertiser, can be improved greatly, which illustrates that the increasing of market segmentation granularity in a certain extent can improve the advertising effect.

(2) From the comparisons of the percentage for the 6 cases, we can see that the revenues of each advertiser increase with the increasing of market segmentation granularity in most cases. Thus, increasing of market segmentation granularity is good for all advertisers in the average sense.

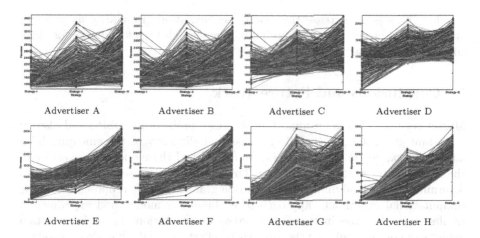

Fig. 5. Comparisons of the revenues for the 8 advertisers under the three strategies in the 2000 experiments

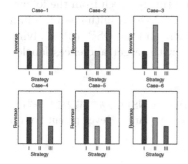

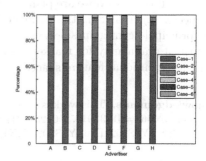

Fig. 6. The 6 cases for the relations of the revenues among the three strategies

Fig. 7. The percentage of each case for the 8 advertisers in the 2000 experiments

4   Managerial Insights

Our paper can offer critical managerial insights for DSPs in RTB markets.

On one hand, there are typically lots of advertisers on the DSP, and how to measure the advertising effect considering all the advertisers is a difficult task for the DSP. This work indicates that simply taking the total revenues of all the advertisers as the optimization objective is appropriate, effective and easy to realize for the DSP.

On the other hand, this work provides a feasible approach to validate the central role of market segmentation strategy on the targeting accuracy and marketing effect for the advertisers in RTB markets. Moreover, given all the optional market segment granularities, this work offers an effective method for DSPs to choose the optimal market segmentation granularity. An easier and practical way for DSPs is to increase the market segmentation granularity as possible, under the condition that there are remaining enough target audience for these advertisers after market segmentation.

5   Conclusions and Future Work

Market segmentation is an important strategic problem for DSPs, and also an important guarantee for the effectiveness and efficiency of the emerging RTB advertising paradigm. In this paper, we established the model for the choice of the market segmentation granularity. This model can be used to help DSPs to determine the optimal market segmentation granularity. We proposed to use the computational experiment approach to evaluate our model, and experimental results show that the increasing of market segmentation degree in a certain extent can not only improve the advertising effect for DSPs, but also is good for all the advertisers.

This work is the first attempt to discuss the RTB market segmentation problem. In our future work, we are planning to extend this research from the following aspects: (a) Analyzing the strategic behavior and the resulting equilibrium of the principal-agent games between the advertisers and the DSP; (b) Studying the optimal granularity for market segmentation with target audiences labeled in a high-dimensional feature space.

Acknowledgements. This work is partially supported by NSFC (#71472174, #7110 2117, #71232006, #61533019, #61233001) and the Early Career Development Award of SKLMCCS (Y3S9021F36, Y3S9021F2K).

References

1. Chen, Y., Berkhin, P., Anderson, B., et al.: Real-time bidding algorithms for performance-based display ad allocation. In: 17th ACM Conference on Knowledge Discovery and Data Mining, pp. 1307–1315. ACM Press, SanDiego (2011)
2. Ghosh, A., Rubinstein, B.I.P., Vassilvitskii, S., et al.: Adaptive bidding for display advertising. In: 18th International Conference on World Wide Web. ACM Press, Madrid (2009)
3. Li, X., Guan, D.: Programmatic buying bidding strategies with win rate and winning price estimation in real time mobile advertising. In: 18th Pacific-Asia Conference on Knowledge Discovery and Data Mining, pp. 447–460 (2014)
4. Wang, F.Y.: Artificial societies, computational experiments, and parallel systems: a discussion on computational theory of complex social-economic systems. Complex Syst. Complex. Sci. **1**(4), 25–35 (2004)
5. Zhang, W., Yuan, W., Wang, W., Shen, X.: Real-time bidding benchmarking with iPinYou dataset. UCL Technical report (2014)
6. Zhang, C., Zhang, E.: Optimized bidding algorithm of real time bidding in online ads auction. In: 21th International Conference on Management Science and Engineering, pp. 33–42 (2014)

Combining Feature-Based and Instance-Based Transfer Learning Approaches for Cross-Domain Hedge Detection with Multiple Sources

Huiwei Zhou[✉], Huan Yang, Long Chen, Zhenwei Liu, Jianjun Ma, and Degen Huang

School of Computer Science and Technology, Dalian University of Technology, Dalian 116024, Liaoning, China
{zhouhuiwei,majian,huangdg}@dlut.edu.cn,
{yanghuan_dlut,chenlong.415,
liuzhenwei}@mail.dlut.edu.cn

Abstract. The difference of hedge cue distributions in various domains makes the domain-specific detectors difficult to extend to other domains. To make full use of out-of-domain data to adapt to a new domain and minimize annotation costs, we propose a novel cross-domain hedge detection approach called FIMultiSource, which combines instance-based and feature-based transfer learning approaches to make full use of multiple sources. Experiments carried on BioScope, WikiWeasel, and FactBank corpora show that our approach works well for cross-domain uncertainty recognition and always improves the detection performance compared to other state-of-the-art instance-based and feature-based transfer learning approaches.

Keywords: Hedge detection · Cross-domain · Transfer learning

1 Introduction

Hedge detection has attracted more and more interest in natural language processing (NLP) community. The CoNLL-2010 Shared Task [1] addresses the detection of uncertainty in two domains (biological publications and Wikipedia articles). Hedging is widely used in various domains from science to humanities [2]. To date, several corpora annotated for uncertainty are publicly available in different domains, such as BioScope [3] from biomedical domain, FactBank [4] from newswire domain, WikiWeasel [1] from encyclopedia domain, etc. The uncertainty cue vocabulary and the distribution of certain and uncertain senses of cues vary in different domains [2].

The currently existing approaches focus on learning the domain-specific detectors in each publicly available corpus. Such approaches works well in the domains in which the training data are sufficient. Though the several publicly available corpora cover different aspect of uncertainty, the domain dependency of hedge detection makes the model trained on existing corpora does not perform well in other domains. It is impractical to annotate training data for each of the application domains. Szarvas et al. [2] show that

© Springer Science+Business Media Singapore 2015
X. Zhang et al. (Eds.): SMP 2015, CCIS 568, pp. 225–232, 2015.
DOI: 10.1007/978-981-10-0080-5_22

distant domain data can contribute to the recognition of uncertainty cues, which efficiently reduces the annotation costs for a new domain. However, the target training data Szarvas et al. [2] need are 1,000 ∼ 2,000 labeled sentences, which are also too expensive.

This paper addresses the problem of detecting hedge in the case that the training data in the target domain are rarely scarce, for example, 200 instances. We propose a robust cross-domain hedge detection method, which combines feature-based transfer learning (or feature-transfer) and instance-based transfer learning (or instance-transfer) approaches to transfer hedge detection knowledge from multiple sources. The proposed method first trains a feature-transfer classifier and an instance-transfer classifier by using each source data individually in combination with the target data. The final ensemble classifier is given by a sum of the individual classifiers' predictions to overcome the weakness of a single transfer approach from a single source. The experiments show that our method outperforms the state-of-the-art transfer learning approaches in hedge detection, especially for a very small insufficient training set.

2 Related Work

The early hedge detection systems use handcrafted cue lexicons for cue recognition [5]. However, not all occurrences of the cues indicate uncertainty. With the development of publicly available corpora, the hedge detection task is treated as a sequential labeling [6, 7] or a token classification problem [8]. The former predicts for all tokens whether a token is inside a cue or outside a cue. The latter matches cue candidates based on a cue lexicon, and then classifies whether they denote uncertainty. All of these studies focus on in-domain cue detection, which assume that the training data are sufficient to get an acceptable model in the application domain. However, there are only several available corpora constructed for special domains, and the set of cues used and frequency of their certain and uncertain usages are domain dependent.

In the case that the training and working data follow a different data distribution, transfer learning could improve the performance without much expensive data-annotation effort. The existing transfer learning approaches can be divided into two cases: feature-transfer learning, and instance-transfer learning approaches. Feature-transfer learning [9, 10] tries to discovery latent common features shared by the source and target domains. Daumé III [9] proposes a Frustratingly Easy Domain Adaptation approach (Hereafter, FruDA for short), which augments the feature space of both the source and target data and uses the result as the input to a standard learning algorithm.

Instance-transfer learning [11–13] attempts to select the important source instances which are reused in the target domain by reweighting. TrAdaBoost [11] focuses on using a large set of labeled source data and a small set of labeled target data to automatically adjust the weights of training instances by Boosting. However, the limited labeled target data cannot represent the whole target domain sufficiently. Active learning algorithms are usually used to choose the target instances for representing the target distribution [12]. On the other hand, TPTSVM [13] uses semi-supervised learning to improve the generalization performance.

Transfer learning relying on one source could lead to negative transfer. Yao and Doretto [14] propose the MutiSourceTrAdaBoost algorithm to import knowledge from multiple sources to decrease the risk of negative transfer. Eaton and desJardins [15] propose a TransferBoost algorithm, which increases the weight of sources that show positive transfer to the target. Dredze et al. [16] adapt multiple source domain classifiers to a new target domain. However, they train domain-specific classifiers individually in each source domain without exploiting a small target training data set.

To take advantages of both feature-transfer learning and instance-transfer learning approaches and import knowledge from multiple sources, this paper proposes a method to integrate the two approaches for cross-domain hedge detection with multiple sources (FIMultiSource). The proposed method first trains a feature-transfer classifier and an instance-transfer classifier by using each source domain data set and a small number of target domain data set. Finally, the classifiers trained in different approaches with different sources are combined to take advantage of their various strengths.

3 Cross-Domain Hedge Detection

3.1 Hedge Cue Detection

Hedge information almost presents by lexical cues. However, the words in lexicon do not always present uncertainty but according to the context. This paper formulates the cue detection problem as a classification problem of candidate cues in lexicon. How to classify the candidate cues by using the limited labeled data in target domain to leverage multiple sources is the main task we want to solve in this paper. Several basic features, which are provided by GENIA Tagger, are used in our system.

(1) Current token features: $Token(i)$ $(i = 0)$
(2) Stem features: $Stem(i)$ $(i = -1, 0, +1)$
(3) POS features: $POS(i)$ $(i = -1, 0, +1)$
(4) Chunk features: $Chunk(i)$ $(i = -1, 0, +1)$
(5) Co-occurrence features: $Co(i)$ $(i = -1, 0, +1)$.

Other candidate cues that occur in the same sentence. Where $Co(-1)$ is the first candidate cue to the left, $Co(+1)$ is the first candidate cue to the right.

3.2 Transfer Learning with Multiple Sources

Given three kinds of data sets, a very small target training set $D_l = \{(x_i^l, y_i^l)|i = 1, \ldots, n\}$, $y_i^l = \{-1, +1\}$, the target domain unlabeled set $D_u = \{(x_j^u)|j = 1, \ldots, m\}$, and multiple source domain labeled sets $D_{s1}, \ldots, D_{st}, \ldots, D_{sq}$, where $D_{st} = \{(x_k^{st}, y_k^{st})|k = 1, \ldots, |st|\}$, $y_k^{st} = \{-1, +1\}$. n, m and $|st|$ are the sizes of D_l, D_u and D_{st}. Assume that D_l and D_u are in the same domain with same distribution, but D_l is not sufficient to training a model to classify the D_u. Our transfer learning algorithm tries to use D_{st} and D_u to help the insufficient training set D_l to train a better classifier $F(x_j)$ that minimizes the prediction error on the target unlabeled set D_u.

We combine instance-transfer and feature-transfer learning approaches to make full use of multiple sources. A formal description of FIMultiSource is given in Algorithm 1. For each source data D_{st}, FIMultiSource trains a feature-transfer classifier on $D_{st} \cup D_l$ and an instance-transfer classifier on $D_{st} \cup D_l \cup D_u$(or $D_{st} \cup D_l$). The final ensemble classifier is given by a sum of the individual classifiers' predictions to overcome the weakness of a single transfer approach from a single source.

Algorithm 1.

Input the source domains labeled data sets $D_{s1}, \ldots, D_{st}, \ldots, D_{sq}$, the target domain labeled data set D_l, the unlabeled target data set D_u, and the number of iteration N.

For $t = 1, \ldots, |sq|$

1. Train a feature-transfer classifier $Feature_t(x)$ on $D_{st} \cup D_l$.

2. Train an instance-transfer classifier $Instance_t(x)$ on $D_{st} \cup D_l \cup D_u$ (or $D_{st} \cup D_l$),
   for $d = 1, \ldots, N$.

end for

Output the final ensemble classifier $F(x_j) = sign(\sum_{t=1}^{|sq|} Feature_t(x_j) + Instance_t(x_j))$.

For feature-transfer, FruDA [9] is adopted for our FIMultiSource method. FruDA [9] projects the source data and target data $X = R^F$ to an augmented space $\tilde{X} = R^{3F}$ by the mappings Φ^s and Φ^t respectively:

$$\Phi^s(x) = <x, x, 0> , \ \Phi^t(x) = <x, 0, x> \tag{1}$$

where, $0 = <0, \ldots, 0> \in R^F$ is the zero vector, F is the dimension of the initial input. Augment reduces difference between a source and a target domain by mapping the source data and target data into a common space. However, the scarce training data in the target domain is insufficient for represent the feature distribution of target domain.

For instance-transfer, TrAdaboost [11] and TPTSVM [13] are used for our FIMultiSource method respectively. Comparing with TrAdaboost, TPTSVM queries the most confident target domain unlabeled data D_u into training data, which helps to overcome the over-fitting of TrAdaBoost.

4 Experiments

4.1 Performance Based on Transfer Learning with One Source

We evaluate our method on three corpora: BioScope [3] from biomedical domain, WikiWeasel [1] from encyclopedia domain, and FactBank [4] from newswire domain. The BioScope corpus contains clinical texts as well as biomedical texts from full paper and scientific abstracts. Only scientific abstract is used as biomedical data in our experiments. The number of cues in biomedical, encyclopedia and news domains is

2694, 3265 and 720 respectively. The number of candidate cues in the three domains is 10272, 17009 and 2758 respectively. The training data in the target domain used in the experiments are only 200 instances.

We first evaluate our feature-transfer and instance-transfer learning combination approach in hedge detection by using only one source domain. Our transfer learning combination method is compared with the following basic methods:

Sou: This method applies the SVM with only the current source training set D_{st}.

Tar: This method applies the SVM with only the target training set D_l.

Sou + Tar: This method applies the SVM with both the source and target training sets $D_{st} \cup D_l$.

Fru: This method applies the feature-transfer learning approach FruDA with both the source and target training sets $D_{st} \cup D_l$.

TrA: This method applies the instance-transfer learning approach TrAdaBoost with both the source and target training sets $D_{st} \cup D_l$.

TPT: This method applies the instance-transfer learning approach TPTSVM with the three data sets: the source training set, the target training set and the target unlabeled set $D_{st} \cup D_l \cup D_u$.

Note that the former three basic methods are the traditional learning methods, while the later three basic methods are the transfer learning methods. To illustrate the effects of our transfer learning combination method, the two instance-transfer approaches (**TrA** and **TPT**) are combined with the feature-transfer approach respectively:

Fru + TrA: This method combines the outputs of the basic methods **Fru** and **TrA**.

Fru + TPT: This method combines the outputs of the basic methods **Fru** and **TPT**.

We generate six groups of experiments using all three data sets. Iteration time N of both TrAdaBoost and TPTSVM is 100. All the results below are the average of 10 repeats by random. The performance for hedge detection is evaluated by the official tool of the CoNLL-2010 shared task. The evaluation for hedge detection is carried out on the cue-level. The cue-level scores are based on the exact match of cue phrases.

The comparison of cue-level F-scores is shown in Table 1. Seen from the table, the proposed feature-transfer and instance-transfer combination approach outperforms the

Table 1. Comparison of Cue-level F-scores with One Source (abc = scientific abstract in biomedical domain; enc = encyclopedia)

Domain		Method							
Target	Source	Sou	Tar	Sou + Tar	Fru	TrA	TPT	Fru + TrA	Fru + TPT
news	abs	51.13	36.83	55.57	63.96	64.17	65.53	65.47	**66.10**
news	enc	67.18	36.83	67.29	68.52	66.24	66.67	68.67	**68.92**
abs	news	71.93	72.17	**84.70**	81.37	83.50	84.25	83.51	84.22
abs	enc	79.40	72.17	80.76	82.27	82.42	82.68	**83.90**	83.80
enc	abs	61.29	28.99	62.27	66.25	70.31	70.84	70.67	**71.19**
enc	news	70.99	28.99	72.08	69.16	72.23	73.23	72.81	**73.97**
Average		66.99	46.00	70.45	71.92	73.14	73.87	74.17	**74.70**

basic methods generally. Among the six basic methods, the three transfer learning methods (**Fru**, **TrA**, and **TPT**) are obviously better than the three traditional learning methods (**Sou**, **Tar**, and **Sou + Tar**). Since the training data are not sufficient (200 instances), the performance of **Tar** is very poor. The average cue-level F-score is only 46.00 %. **Sou + Tar** could improve **Tar** with the help of the source data. For the transfer learning methods, **TPT** always improves the F-scores of **TrA** by using the target unlabeled data to help transfer learning. Overall, **Fru + TPT** gives the best cue-level F-score 74.70 % through the novel combination of feature-transfer and instance-transfer learning as well as the utilization of the target unlabeled data.

4.2   Performance Based on Transfer Learning with Multiple Sources

We then perform the experiments with multiple sources. Three groups of experiments are generated using the same three data sets. Our FIMultiSource method is compared with the following basic methods employing multiple sources:

MSou: This method applies the SVM with the union of multiple source training sets $D_{s1}\cup\ldots\cup D_{sq}$.

MSou + Tar: This method applies the SVM with the union of multiple source training sets and the target training set $D_{s1}\cup\ldots\cup D_{sq}\cup D_l$.

MFru: This method extends FruDA to multiple source domains with the multiple source training sets and the target training set $D_{s1}\cup\ldots\cup D_{sq}\cup D_l$. For $K=q+1$ domains totally, FruDA is simply extended to multiple sources by expanding the feature space to $R^{(K+1)F}$, where "+1" corresponds to the "general domain" [9].

MTPT: This method extends TPTSVM to multiple source domains with the union of multiple source training sets, the target training set and the target unlabeled set $D_{s1}\cup\ldots\cup D_{sq}\cup D_l\cup D_u$. Zhou et al. [13] introduces a source factor $e^{(\varepsilon_t-\varepsilon_s)}$ to increase or decrease the source weight based on whether the union of the source and target domain data shows positive or negative transfer to the target domain. ε_t is the prediction error of the classifier trained with the target sets. ε_s is the prediction error of the classifier trained with both the target and source sets. We simply extend TPTSVM to multiple sources by using the source factor $e^{(\varepsilon_t-\varepsilon_{st})}$ to adjust the weight of source domain set D_{st}.

ComFru: This method combines the prediction value of each **Fru**, which is learned by transferring knowledge from each source to the same target.

ComTPT: This method combines the prediction value of each **TPT**, which is learned by transferring knowledge from each source to the same target.

The comparison of cue-level F-scores is shown in Table 2. From the table we can see that the proposed **FIMultiSource** method outperforms all six basic methods over all groups. Among the six basic methods, **MSou** and **MSou + Tar** perform poorly because they simply employ traditional learning with the union of training data. The average cue-level F-score of **MFru** is 69.59 %, while that of **MTPT** is 72.38 %. **ComFru** and **ComTPT** which combine the results of individual classifier could perform better than **MFru** and **MTPT**. The average cue-level F-scores of **MFru** and

Table 2. Comparison of Cue-level F-scores with Multiple Sources

Domain		Method						
Target	Source	MSou	MSou + Tar	MFru	MTPT	EnsFru	EnsTPT	FIMultiSource
news	abs, enc	55.00	57.16	62.18	64.35	68.27	67.71	**69.23**
abs	enc, news	78.94	80.14	81.48	82.09	82.93	84.75	**84.96**
enc	abs, news	61.16	61.78	65.10	70.71	69.14	73.54	**73.61**
Average		65.03	66.36	69.59	72.38	73.45	75.33	**75.93**

MTPT are even lower than that of **Fru** and **TPT** (see Table 1). **MFru** and **MTPT** adapt multiple sources to a new target domain in the transfer learning process. However, they do not get the results as good as expected. This is perhaps because transferring hedge detection knowledge from multiple sources to the target domain in one task is quite difficult. Our **FIMultiSource** achieves the cue-level F-score of 75.93 %, which is 1.23 % higher than that of **Fru + TPT**. The improvements benefit from multiple sources.

Szarvas et al. [2] uses the same three corpora (Bioscope, WikiWeasel, and FactBank). The target training set they need consists of 1,000 ∼ 2,000 sentences. The model trained by these sentences has already achieved the average F-score of 73.5 %. Szarvas et al. [2] adapt FruDA [9] to transfer hedge detection knowledge from one source to the target domain, and the average F-score reaches 77.8 %. We also employ FruDA in our experiments, and achieve the average F-score of 71.92 %. However, the target training data we need is rarely scarce (200 instances). By using our FIMultiSource method, the average F-score is improved to 75.93 %, which outperforms FruDA significantly.

5 Conclusion

In this paper, we proposed a novel approach FIMultiSource for cross-domain hedge detection with multiple sources. The proposed approach first learns cross-domain classifiers from each source to the target domain based on the feature-transfer and instance-transfer learning approach respectively. The final ensemble classifier is given by a sum of the individual classifiers' predictions to overcome the weakness of a single transfer approach from a single source. Experiments show that our FIMultiSource achieves the cue-level F-score of 75.93 %, which significantly outperforms previous state-of-the-art transfer learning approaches. Both feature-transfer and instance-transfer play an important role in cross-domain hedge detection. How to balance their function to optimize the combination of feature-transfer and instance-transfer for cross-domain hedge detection will be studied in our future work.

Acknowledgements. This research is supported by National Natural Science Foundation of China (Grant No. 61272375).

References

1. Farkas, R., Vincze, V., Móra, G., Csirik, J., Szarvas G.: The CoNLL-2010 shared task: learning to detect hedges and their scope in natural language text. In: Proceedings of the 14th Conference on Natural Language Learning (CoNLL-2010) Shared Task, pp. 1–12 (2010)
2. Szarvas, G., Vincze, V., Farkas, R., Móra, G., Gurevych, I.: Cross-genre and cross-domain detection of semantic uncertainty. Comput. Linguist. **38**, 335–367 (2012)
3. Vincze, V., Szarvas, G., Farkas, R., Móra, G., Csirik, J.: The BioScope corpus: biomedical texts annotated for uncertainty, negation and their scopes. BMC Bioinf. **9**, S9 (2008)
4. Saurí, R., Pustejovsky, J.: FactBank: A corpus annotated with event factuality. Lang. Resour. Eval. **43**, 227–268 (2009)
5. Light, M., Qiu, X.Y., Srinivasan, P.: The language of bioscience: facts, speculations, and statements in between. In: Proceedings of BioLink 2004 Workshop on Linking Biological Literature, Ontologies and Databases: Tools for Users, pp. 17–24 (2004)
6. Tang, B.Z., Wang, X.L., Wang, X., Yuan, B., Fan, S.X.: A cascade method for detecting hedges and their scope in natural language text. In: Proceedings of the 14th Conference on Computational Natural Language Learning: Shared Task, pp. 13–17 (2010)
7. Zhou, H.W., Li, X.Y., Huang, D.G., Yang, Y.S., Ren, F.J.: Voting based ensemble classifiers to detect hedges and theirs scopes in biomedical texts. IEICE Trans. Inf. Syst. **E94-D**(10), 1989–1997 (2011)
8. Velldal, E., Øvrelid, L., Oepen, S.: Resolving speculation: MaxEnt cue classification and dependency-based scope rules. In: Proceedings of the 14th Conference on Computational Natural Language Learning: Shared Task, pp. 48–55 (2010)
9. Daumé III, H.: Frustratingly easy domain adaptation. In: Proceedings of the 45th Annual Meeting of the Association of Computational Linguistics, pp. 256–263 (2007)
10. Zhang, H.N., Tian, Z., Kuang, R.: Transfer learning across cancers on DNA copy number variation analysis. In: Proceedings of the 13th International Conference on Data Mining (ICDM), pp. 1283–1288 (2013)
11. Dai, W.Y., Yang, Q., Xue, G.R., Yu, Y.: Boosting for transfer learning. In: Proceedings of the 24th International Conference on Machine learning, pp. 193–200 (2007)
12. Wang, X.Z., Huang, T.K., Schneider, J.: Active transfer learning under model shift. In: Proceedings of the 31st International Conference on Machine Learning (ICML-2014), pp. 1305–1313 (2014)
13. Zhou, H.W., Zhang, Y., Huang, D.G., Li, L.S.: Semi-supervised learning with transfer learning. In: Proceedings of the Chinese Computational Linguistics and Natural Language Processing Based on Naturally Annotated Big Data, pp. 109–119 (2013)
14. Yao, Y., Doretto, G.: Boosting for transfer learning with multiple sources. In: Proceedings of the 2010 IEEE Conference on Computer Vision and Pattern Recognition (CVPR), pp. 1855–1862 (2010)
15. Eaton, E., desJardins, M.: Selective transfer between learning tasks using task-based boosting. In: Proceedings of the 25th AAAI Conference on Artificial Intelligence, pp. 337–342 (2011)
16. Dredze, M., Kulesza, A., Crammer, K.: Multi-domain learning by confidence-weighted parameter combination. Mach. Learn. **79**(1–2), 123–149 (2010)

A Music Recommendation Algorithm Based on Hybrid Collaborative Filtering Technique

Yan Yan, Tianlong Liu, and Zhenyu Wang[✉]

School of Software Engineering, South China University of Technology, Guangzhou, People's Republic of China
{gwyanyan,wangzy}@scut.edu.cn

Abstract. With the rapid development of e-commerce, the recommendation system is becoming increasingly important. The collaborative filtering algorithm is one of the most successful algorithms in recommendation system. However, users' preference and application scenarios play a significant role in determining the effect of these algorithms. Therefore, we propose a method to fusion item-based and user-based collaborative filtering algorithms into a hybrid one, we also consider time as a property and design a method to evaluate the rank of items. Last, the Movielens and real-world data-set are used to evaluate the recall and efficient of the algorithm in this paper.

Keywords: Collaborative filtering · Recommendation systems · Ensemble method · Time-awareness

1 Introduction

With the rapid development of digital music, personalized music recommendation has gradually attracted the attention of researchers, and many remarkable research achievements were produced. Many popular online digital music radios, such as Pandora, Last.fm, Song Taste and Douban, have already adopted different kinds of recommendation techniques. However, the accuracy and the coverage ratio of their recommendation correspondences are not satisfactory, and be lack of personalization.

In the field of personalized recommendation, many approaches are single recommendation based. For instance, Pandora takes advantage of expert-based recommendation; Song Taste adopts the content-based one. Lee S. [1] applies the random-walk method to recommend music by utilizing the graph mode composed by user and music. However, other content-based recommendation algorithms are more use of audio analysis technology to recommend [2–5]; as for additional researches, the hybrid model is adopted, such as Yoshii K established a hybrid recommendation model which combined with content-based and collaborative filtering algorithms, while Tso-Sutter combined the two kinds of collaborative filtering algorithm with label features. About these algorithms mentioned above, collaborative filtering is widely adopted because of its simple ideas and stable performance. The collaborative filtering techniques can be classified into two categories: model-based and memory-based, which basic idea is to predict the behavior of a person who has similar preferences or habits.

© Springer Science+Business Media Singapore 2015
X. Zhang et al. (Eds.): SMP 2015, CCIS 568, pp. 233–240, 2015.
DOI: 10.1007/978-981-10-0080-5_23

After analyzing the traditional collaborative filtering algorithm and the existing problems, this paper proposed a recommendation system framework which combined two different collaborative filtering algorithms. And make use of time and weight penalty strategy to resolve popular items problem in practical applications. Experimental results indicate that the new proposed method for the top-N recommending issue can obtain more accurate and better performance than single collaborative filtering algorithm or other hybrid algorithms.

2   Related Work

In this paper, $U \times I \rightarrow R$ represent the user-item scoring matrix, where U indicates the user set, and I as the item set. Besides, r(u, i) represents the score value of the user u to the item i in the score matrix R, where $r \in R$, $u \in U$, $i \in I$.

Different from content-based recommendation algorithm, the similarity between items is not calculated by item features, but by analyzing the score of these similar items which are interested to users [6]. The content-based collaborative filtering algorithm has been widely used in industry, and the most famous one is the item-to-item technique which proposed by Amazon [7]. Another one is user-based approach which applied by Grouplens [8]. In practice, data are so sparse that many recommendation techniques focus on solving the issue of data sparsely. Yan et al. [9] propose a method using two weighted scoring techniques to fill the scoring matrix, and utilize the score matrix to recommend. This method solves the problem of matrix sparseness in some extent. Wang et al. [10] utilized the technique of k-means to gather different items together to represent a specified type of interest. A user may have a variety of hobbies, and it can be handled in a different interest area, according to the actual demand when recommending, they also introduce a weight coefficient of items to reduce the influence of similarity when two users preferred common popular items. For the issue of data sparseness may influence the recommendation, the similarity transfer algorithm is proposed by Xie [11]. According to a specified threshold, this algorithm is used to find the misaligned similarity value, and then replace it with transfer similarity value. The extent of transfer is defined by the degree of interaction, which is expressed as the amount of common items between user U_i and U_j. Thus, the original similarity is adopted when the user's interaction degree is smaller than threshold.

3   Fuse the Collaborative Filtering Algorithm

3.1   Algorithm Overview

The fusion of two kinds of collaborative filtering algorithm is the continuation and development for the traditional collaborative filtering approach. The approach that based on the item can apply to more users. As long as the user has an item evaluation, the similar unevaluated items can be recommended to user. Besides, the Item-CF have a strong ability to explore the long tail item, which can not only fully reflect the users' personalized demands, but also can provide a convincing explanation. In contrast, although User-CF has no credible explanation provided for recommended result and nor in the absence of user behavior to recommend, it can works well when the items amount far greater than users and it also have a strong real-time advantage. By combining these two algorithms, the results of the recommendation can be combined in a linear way, which can reduce the disadvantages to an extent of single approach in different applications. The strategy mentioned above not only can make a credible explanation for recommendation results, but also can obtain the stronger robustness for data processing, which can be adopted to different size of user sets and items and can balance the single algorithm's disadvantages in timeliness and diversity.

3.2 Algorithm Design

The fusion framework for collaborative filtering is based on the idea of memory-based, which is mainly divided into two phases, the first phase is preparing data, computing user-item matrix and calculating the similar user set and similar item set by combining the user evaluating score. Moreover, the second phase is to generate the top-N recommendation by using the scoring matrix and similar sets, and then improving and fusing the recommended results.

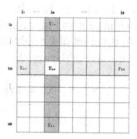

Fig. 1. User-item scoring matrix

As shown in Fig. 1, we first construct a scoring matrix for original user-item data. Then, for any user u_i, we use the following formula to calculate the Pearson correlation coefficient of similarity:

$$\text{sim}\left(u, u'\right) = \frac{\sum_{i \in I_{u,u'}} (R_{u,i} - \overline{R_u})(R_{u',j} - \overline{R_{u'}})}{\sqrt{\sum_{i \in I_{u,u'}} (R_{u,i} - \overline{R_u})^2} \sqrt{\sum_{i \in I_{u,u'}} (R_{u',i} - \overline{R_{u'}})^2}} \tag{1}$$

Where, $I_{u,u'}$ are the common items evaluated by user u and u', $R_{u,i}$ represents the evaluation made by user u to item i, $\overline{R_u}$ is the average evaluation score of items evaluated by user u, finally the similarity between user u and u' is calculated. In Item-CF, the calculation of items' similarity according to the similarity of users who evaluated items, so it ignores a single Item grading. Because we focus on Top-N recommendations, so we used the similarity of Loglikelihood [12] in Mahout [13]:

$$\text{sim}\left(i, i^{'}\right) = 1 - 1/ \begin{array}{l} 1 + 2 \times [Entropy\left(C_{ii'} + C_{i'}, C_i + C'\right) \\ + Entropy\left(C_{ii'} + C_i, C_{i'} + C'\right) \\ + Entropy\left(C_{ii'}, C_{i'}, C_{i'}, C'\right) \end{array} \tag{2}$$

Where *Entropy* is used to calculate the Shannon Entropy, C_{ii} is the number of users who have evaluated on item i and i', $C_{i\cdot}$ is the number of users who have only evaluated on item i', and C_i is defined similarly, C' is representative of the number of users who have not evaluated on item i or i'. The similarity of Loglikelihood distribution in [0,1]. The user's Top-N recommendation list can be generated after computing the nearest neighbor of users and items. In User-Based algorithm, the recommended level of user u on item i can be calculated by counting the item's frequency of occurrence on the similar users and users' similarity.

$$p^{user}\left(u, i\right) = \frac{\sum_{v \in N_u \cap O_{v,i}=1} \text{sim}(u, v)}{|N_u|} \tag{3}$$

$\text{sim}(u,v)$ is the similarity of users u and v, N_u is the set of k neighbors who have most similarity to user u, which k is determined by the size of dataset, and $O_{v,i}$ indicates that user v evaluated item i, but user u does not evaluate it.

In Item-CF algorithm, the weight of item i in recommendation list is calculated by all users who evaluated item i, N_i is the set of k neighbors who have most similarity to item i, which k is determined by the size of dataset, the formula is defined as follows:

$$p^{item}\left(u, i\right) = \frac{\left|\{v \in N_u | O_{v,i} = 1\}\right|}{|N_i|} \tag{4}$$

After calculated weight using the two formulas above, we can get two recommendation lists for user u. Because of the User-CF and Item-CF need normalized score, because each algorithm has different ways of scoring. The new formula of fusion way is as follows:

$$p(u, i) = \theta \times \frac{p^{user}}{\sum_u p^{user}} + (1 - \theta) \times \frac{p^{item}}{\sum_i p^{item}}. \tag{5}$$

3.3 Weight Penalty on the Importance of Time and Item

"Harry Potter" problem has often appeared in collaborative filtering algorithms, the most popular items have a higher rating by users regularly, so the personalized recommendation system cannot be established. For this purpose, we use time information to attenuate popularity in the similarity calculation. And when the Top-N recommendation list is generating, we also use a punishment mechanism to reduce the influence of popularity of items. The track of user's action is affected by their personal interests and their neighbors, and time is also an important factor. The user's interests are changed with time. For example, the programmers just like the introduction of programming technology books when they are rookies, but they will be interested in more in-depth and professional books as time goes by. Therefore, if we want to accurately predict the user's current interests, we must consider the influence of time on user's interests.

Without considering the effects of time on user's interest, the evaluation $r_{u,i}$ is the user's preference of each item and it can be obtained from dataset. So we use an evaluation timestamp t_r when user u evaluated item i, under the condition of the current timestamp t, the evaluation based on time is formulated as follow:

$$r^t_{u,i} = r_{u,i} \times \frac{1}{1 + \log\left(t - t_r + 1\right)} \tag{6}$$

In different times, people's interests are affected by many factors, leading to the importance of items changed. Thus, we focused on the weight penalty for popular items, in order to reduce the impact of "Harry Potter" problem. First of all we must statistic the number U of users, then for each item, the set of users u_i who are interested to items is needed. For each item i, the importance in the current time R^t_i can be indicated as:

$$R^t_i = \left(U/u_i\right)^\alpha \tag{7}$$

Thus, the more people concern about it, the less weight in the personalized recommended list. And more unpopular the items, the higher weight in the calculation. When the final recommendation list is generated combining with the importance of the items, we need to correct the Eq. (7) for added importance weight of item i:

$$P^t_{u,i} = p(u, i) \times R^t_i. \tag{8}$$

4   The Experiment

4.1   The Experimental Data

In order to verifying the validity of the proposed method, we adopted Movielens-100 K of "Grouplens" (http://grouplens.org/datasets/movielens/) for a wide range of experiments. "Movielens" data set contains 1,682 movies, 943 users and 100,000 ratings. Since we need to use the time information, so first of all we need to sort each user ratings according to timestamp, then split the data follow the rule of 90 %/10 %, where 90 % as the training set and 10 % as the test set. Data of test set is used to calculate the recall rate, coverage, etc. Each user in test set has 10 items; and the recommendation system recommended 10 items for each user.

In addition, we use the "Imusic" user behavior data of Guangzhou telecom to verify the results on music recommendation. "Imusic" includes 917388 users and 114302 songs, users historical behavior in 4 weeks is extracted. We treated weekly data as a separate set, and picked up last 10 use records as a test set.

4.2 Results

Movielens Data Experiment. As shown in Fig. 2, the first experiment shows the recall changed under different θ values. The highest rate of recall is 22.86 % when θ is 0.8.

Fig. 2. The influence of different parameters on the results

Experiment 2, which result depicted in Fig. 3, revealed the different recall rate when we selected different neighbors of the collaborative filtering fusion algorithm. The highest rate of recall is 23.20 %, when the number of neighbors is 40.

Fig. 3. The influence of different neighbors for recall rate

The third experiment shows the recall rate of different algorithms; it contains the User-CF algorithm, the Item-CF algorithm, the fusion algorithms with or without weight penalty and time factor, and the TST algorithm which we introduced in Sect. 2. According to the result of Fig. 4, we can see that the fusion algorithm with time and weight optimization has the highest recall rate of the Top-N recommend problem.

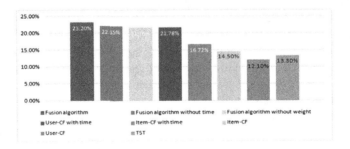

Fig. 4. Different recall rate of algorithm

Imusic Data Experiment. The experiment based on real users' behaviors of "Imusic" focus on three aspects: the hits rate, the recall rate and the coverage rate. We tested three algorithms, the Item-CF, the optimized and unoptimized version of the fusion collaborative filtering algorithm. The result can be seen from the Fig. 5, that the fusion algorithm can predict 90 % users' interests. That means there is at least one song in the recommended list which is listened in the test set. On the other hand, simply use the Item-CF algorithm the hits rate is only 53 %. As shown in Figs. 6 and 7, the results of recall rate and song coverage rate also showed that the advantage of fusion algorithm. Besides, the optimized algorithm using time and weight penalty is better than the unoptimized one.

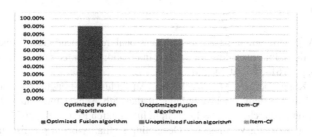

Fig. 5. Hits rate comparison of different algorithms

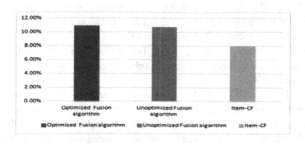

Fig. 6. Recall rate comparison of different algorithms

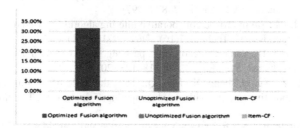

Fig. 7. Coverage rate comparison of different algorithms

5 Conclusion

This paper proposed a personalized recommendation algorithm which integrated two collaborative filtering algorithms. And after we optimized the algorithm with the importance of items and time factor, it can be more accurate for users to recommend items consistent with their interests. The experiment shown that the optimized algorithm using time and weight penalty has a better correspondence than other algorithms, and there is a satisfactory result on Top-N recommendation.

References

1. Lee, S.: A generic graph-based multidimensional recommendation framework and its implementations. In: Proceedings of the 21st International Conference Companion on World Wide Web, pp. 161–166. ACM (2012)
2. Cano, P., Koppenberger, M., Wack, N.: An industrial-strength content-based music recommendation system. In: Proceedings of the 28th Annual International ACM SIGIR Conference on Research and Development in Information Retrieval, p. 673. ACM (2005)
3. Pohle, T., Pampalk, E., Widmer, G.: Generating similarity-based playlists using traveling salesman algorithms. In: Proceedings of the 8th International Conference on Digital Audio Effects (DAFx 2005), pp. 220–225 (2005)
4. Kuo, F.F., Shan, M.K.A.: Personalized music filtering system based on melody style classification. In: Proceedings of the 2002 IEEE International Conference on Data Mining, pp. 649–652. IEEE Conference Publication (2002)
5. Sotiropoulos, D.N., Lampropoulos, A.S., Tsihrintzis, G.A.: MUSIPER: a system for modeling music similarity perception based on objective feature subset selection. User Model. User-Adap. Inter. **18**(4), 315–348 (2008)
6. Breese, J.S., Heckerman, D., Kadie, C.: Empirical analysis of predictive algorithms for collaborative filtering. In: Proceedings of the Fourteenth Conference on Uncertainty in Artificial Intelligence, pp. 43–52. Morgan Kaufmann Publishers Inc. (1998)
7. Linden, G., Smith, B., York, J.: Amazon. com recommendations: Item-to-item collaborative filtering. Internet Comput. **7**(1), 76–80 (2003). IEEE
8. Konstan, J.A., Miller, B.N., Maltz, D., et al.: GroupLens: applying collaborative filtering to Usenet news. Commun. ACM **40**(3), 77–87 (1997)
9. Yan, Z., Shi, L.-H.: An improved collaborative filtering algorithm based on combining user with item. Comput. Knowl. Technol. **7**(16), 3969–3971 (2011)
10. Wang, D.-P., Li, Z.-L., Yang, C.: Knowledge push algorithm based on hot item punishment and user interest change. Syst. Eng. **32**(1), 118–123 (2014)
11. Xie, F., Chen, Z., Xu, H., Feng, X., Hou, Q.: TST: threshold based similarity transitivity method in collaborative filtering with cloud computing. Tsinghua Sci. Technol. **03**, 318–327 (2013)
12. Vol, N.: Accurate methods for the statistics of surprise and coincidence. Comput. Linguist **19**, 61–74 (1993)
13. http://mahout.apache.org/

Author Index

Printed in the United States
By Bookmasters

Printed in the United States
By Bookmasters